Geology of the Great Basin

GEOLOGY OF THE

GREAT BASIN

BILL FIERO

RENO: UNIVERSITY OF NEVADA PRESS: 1986

GREAT BASIN SERIES EDITOR: JOHN F. STETTER

Copyright © University of Nevada Press 1986
All Rights Reserved
Printed in the United States of America
Designed by David Comstock

Library of Congress Cataloging-in-Publication Data

Fiero, Bill, 1936–

Geology of the Great Basin.

(Max. C. Fleischmann series in Great Basin natural history)

Bibliography: p.

Includes index.

1. Geology—Great Basin.   2. Mines and mineral resources—Great Basin.   3. Hydrology—Great Basin.   I. Title.   II. Series.

QE79.F54   1986   557.9   85-14001

ISBN 0-87417-083-4 .

ISBN 0-87417-084-2 (pbk.)

Cover photograph: Death Valley. *John Running.*

page i. Hot springs, Black Rock Desert, Nevada. *John Running.*

page ii. The Black Rock. *Tom Brownold.*

page iii. Cathedral Gorge, Nevada. *Tom Brownold.*

title page (iv–v). High Rock Canyon, Nevada. *Tony Diebold.*

pages vi–vii. Great Salt Desert. *John Running.*

page viii. Zabriske Point, Death Valley. *John Running.*

For my family, in recognition of the love and support that allows each of us the confidence to stand alone with our heads tall and with our minds stretched toward the stars. —And for the family of man, in the belief that we will all learn to do the same, in respect and peace—recognizing the common strength we derive from diversity.

Contents

Zabriske Point. *Tom Brownold.*

Preface

WRITING A BOOK on the geology of the Great Basin is like guiding a tour through a great art gallery. Some wish to stop and examine every picture and every detail. Others are anxious to view only the great masterpieces. I decided to describe only what I consider to be the great works of art, and so there is much left unsaid. But without knowledge and appreciation, the pictures might as well be turned to the wall. I hope this book will help reveal some of the beauties of the Great Basin.

The earth does not easily share its secrets. Certainly, no one individual is privy to all the confidences it keeps. Thousands of geologists have been prying around the recesses and alcoves of the Great Basin for a century. The stories and concepts in this book are based on their observations and theories. This book is an attempt to synthesize their ideas for general understanding by nongeologists.

Since the book is written for everyone, the usual method of scientific notation of sources is not used in this volume. The references include what I consider to be the most useful articles on Great Basin geology that will help take the reader deeper into the geological literature. These articles were most heavily relied on for this book. There are also no explicit references in the book to the illustrations used. These have been placed near the relevant text.

Geologists utilize a precise method of capitalization when using the words early, middle, and late to describe geologic ages. The time spans referred to as periods demand a capital, whereas the larger time units, eras, and the smaller units, epochs, are lowercase. I have followed geologic convention in this book.

The geological stories of the Great Basin are too fascinating to remain sequestered in the scientific literature. There they are often too scattered for easy access or are couched in a language undecipherable to nonscientists. The sole purpose of this book is to share the geologic tales that have been slowly and painstakingly pieced together since the first humans traversed this region. I hope they will continue to challenge and intrigue curious humans for millennia to come.

Great Basin rocks, histories, and landscapes are unique. Nowhere else is the combination or the timing exactly the same. Nowhere else can these stories be told. They are the Great Basin.

Railroad Valley from the Grant Range. The center of the
Great Basin. *Stephen Trimble.*

Valley of Fire, Nevada. *Rick Stetter.*

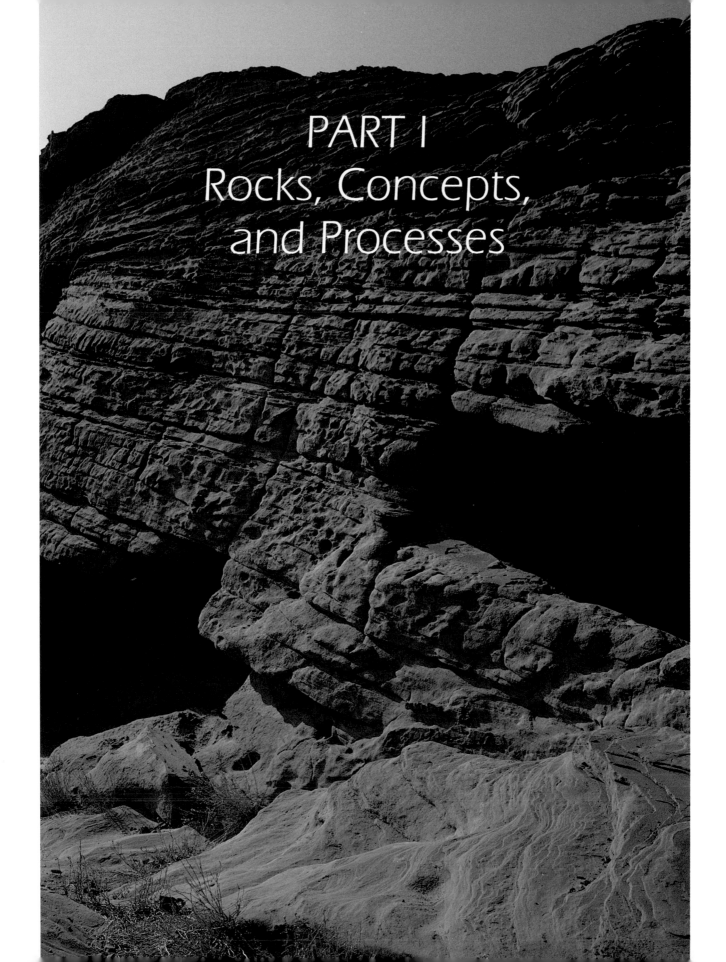

PART I
Rocks, Concepts, and Processes

Black Rock Springs, northern Nevada. These springs
nourished the emigrants after the arduous journey across
the Black Rock Desert. *Tom Brownold*.

1 Introduction

THE DRY SAND stretches away as far as I can see to the east and the west. The July sun burns a hole in the sky and sears the desert basin. Shimmering heat waves ripple the sand at the horizon and make the reflections of the mountains flicker upside down on the pseudoliquid surface. The sandy mud has shrunk and cracked since the last summer rain and has formed polygonal fractures in the barren brown surface. These are the only marks on the sand. One hundred years ago, feet and iron-bound wooden wheels churned the dust of this desolate sandy valley. Emigrants, bound for Oregon, took a shortcut across this forsaken stretch. It was a dash across the sand before the sun parched their tongues black with thirst and shriveled the stock to walking skeletons, or before the sky blackened with the rare but fatal summer rain that would flood the dust-covered basin with water and turn the silts to viscous mud. With wagons mired up to their axles, the emigrants foundered in a sea of muck.

Here, in the Black Rock Desert of what would become northern Nevada, the emigrants would watch for a large dark knob across the shimmering sands and beeline straight for it. Below this black rock and slightly to the north is Double Hot Springs. Here the parched people and stock would find boiling hot water, confirming the Devil's role in fashioning this land. Downstream, where the water had cooled sufficiently to allow the growth of thick mats of grass, they found respite from the drought lands. This is an unforgiving land, a land of rapid change and contrast. This is the Great Basin of the American desert. What geologic forces created this mountain-studded sand- and sage-filled bowl?

Curious, I visited the emigrants' landmark. Turning the black rocks over in my hand, I marveled at the shiny crystalline texture. The rock is an exotic. A foreigner. It is not a resident of the original North American continent, but a piece of another land that crushed against the western edge of the continent several hundred million years ago. The emigrants were orienting their journey by another emigrant. I pondered the irony. How do we know the black rock is not a part of our continent? If pieces of crust move, what propels them? Are their motions controlled by fundamental laws of physics or do they move in a random fashion? I felt like a cerebral emigrant venturing timidly into the unknown of an abysmally deep past.

Miles away, I stood atop the crags that crown a desert mountain range. The cold wind tousled my hair and gently nudged my body toward the cliff edge. Mountains. Through squinted, wind-burned eyes washed with tears, all I could see were mountains. Great walls of mountains leaned against the horizon like serrated stage sets. They tilted in ordered ranks above the sage-green valleys. Essentially north-south. Hundreds of miles long. Shuddered by earthquakes and punctured by volcanos, the ranges stand like ships lying at anchor before a steady north wind. Why?

In the middle of the sagebrush desolation, in the central part of a valley surrounded by snow-capped mountains, steam rose above the winter-browned grasses. In the desert, where water is the most precious and scarce resource, I lay in hot water up to my neck. And I had a choice of not one pool, but many. This area was named Thousand Springs, but I would wager they have never been counted. "Water, water everywhere," here in a land of scarcity. Why?

Sand makes a terrible bed. Its apparent stability is a well-known illusion, and I should have known better. Once again I shake the grains from my bed-

Mono Lake, California. Limey salts from underneath springs create fantastic shapes along the shores of receding Mono Lake. *John Running.*

roll as I have at Jarbidge and Cathedral Canyons and in valleys called Death and Fire. I have tidied my bedroll in countless other canyons, valleys, and ranges, many without name. If one is to study sand, one must learn to live with it.

I am a geologist, and to study the earth I must accept a little grit with the grain. For thirty years I have had an intimate relationship with the earth. We've had our quarrels, inconsequential like the sand grains or more serious when a cliff face collapses. But these are lovers' squabbles and of little import. We share secrets. Rambling throughout desert lands, poking in hidden recesses, struggling on glacial ice, diving into lakes and ocean, floating in creeks and rivers—it has been a beautiful relationship.

Geologists have a special gift to offer others: a sharing of their intimate and personal experience with the earth. This comradeship permits a special ability to converse with the earth and to decipher its experiences. But it is a kiss-and-tell relationship. Knowing a little of earth's past, one likes to gossip. What makes the basins and ranges? What wears them away? What causes continents to crash blindly together? What causes valleys, sand dunes, mineral deposits, deep canyons? The answers are not complete, since the earth is reluctant to share and our intelligence is primitive. Geologists have slept in many sandy bedrolls and gazed at stars far into the night trying to find answers to these queries. Some of our ideas read more like fiction than fact, and perhaps they are. But, like most campfire tales, they are fun to share.

For years I have been conversing with the earth and teaching others how to ask questions and listen to its responses. In all my travels through distant continents, nowhere are the questions more perplexing or the answers more vague and intriguing than here in the Great Basin.

Geologic processes of uplift and erosion are in a dynamic tug of war here. Mountains are young and still rising. Shorelines of deep lakes etch the valley slopes, now burning with desert drought. Titanic forces have been locked in massive conflict. The inexorable upward pressure from heat imbalances in the earth's interior has been countered by the destructive leveling power of water and ice. The unceasing struggle between these basic elements, fire and water, has created the grandeur of the terrain. These lands are unique landscapes, the result of geologic processes relentlessly struggling for primacy through the millennia. We seek an explanation for the scenery and like detectives we use the clues left behind by the agents of uplift and destruction. To unravel the ancient plot requires another view, a new set of eyes, unencumbered by time or space. So equipped, we may trace the evidence to solve mysteries of the past.

Geology is the youngest science. Perhaps this is because it is so imprecise. Physicists and chemists solve problems to the sixth decimal place on their computers. Biologists poke and dissect in the laboratory, studying a thin slice of time—the present. Geologists walk around on the little of the earth we see and scratch their heads about the rest. They check their watches for lunchtime, interrupting a conversation about events that occurred 375 million years ago, plus or minus 5 million years. Scientists don't like imprecision—it smacks of fiction. Perhaps this is why geology is so young; few scientists dared to deal a deck with so many cards missing.

Geology, imprecise as it often is, still offers us the ability to play Sherlock Holmes and partially decipher the past. Perhaps we even better comprehend the present. A mere lifetime is no longer a restriction. Crossing the barrier of our own span of existence and that of our species, we can expand our horizon backward through time and trace ancient events and dramas. We can mentally visualize the past, and we can therefore place the present into a different perspective. A mountain, canyon, or cliff is no longer merely a scene. The view becomes an exciting mystery story, and we become the investigators seeking answers. This different view of our world, the geologic vision, can alter and enrich the lives of all of us. We need only ask the simple question "Why?" and then use our new vision—our geologic eyes—to see the clues and unravel the mystery.

Valley of Fire, Southern Nevada. The black patina of oxides coats ancient sandstones with desert varnish. The Anasazi, or Ancient Ones, have carved petroglyphs through the varnish. *John Running*.

Fly Geyser, northern Nevada. This land of contrast matches abundance of water against the sere drought-lands. New rocks form into ledges as salts precipitate from the hot waters. *John Running*.

2 The Great Basin

As a child, I would trace the outlines with a finger and follow the borders of earth's great features on maps. My fingertip could scale the Himalayas, struggling upward beside Tenzing Norgay and Sir Edmund Hillary; sail the South Pacific with the great Polynesian voyagers; explore the icy recesses of the Arctic, searching for Sir John Franklin and the crew of the *Erebus*. But nowhere did my finger roam a more mysterious region than the Great Basin. The Great Basin! A youthful imagination conjured visions of a vast sandy bowl, replete with rattlesnakes and alkali flats, rimmed totally by lofty mountains. Nowhere did the map depict a river or even a creek flowing out of the great sink. Everywhere the drainage lines slanted toward the interior. My finger would quiver as it walked the rimrock around one quarter of a million square miles of the dry bowl. The basin contains almost all of the state of Nevada, half of Utah, and portions of Oregon, California, and Idaho. New England would be lost in one corner.

The east and west boundaries were easy to trace. The snow-covered heights of the Sierra Nevada blocked the westward drainage toward the Pacific. From the Klamath area of northern California to the Mojave Desert of southern California, the great diagonal northwest-southeast Sierran wall, the highest range in the continental United States, hemmed in the Great Basin to the west. The eastern rim was as distinctly easy to finger-trace as was the western edge. Another great diagonal wall of upthrust rock—the Wasatch Mountains and the high plateaus of southern Utah—formed north-south barriers to any eastern drainages.

The northern rim, less distinct, ambled east-west along the low lava-capped ridges of the Snake River Plains of Oregon and Idaho. The tributaries of the Deschutes, John Day, Owyhee, and Snake rivers appeared in the atlas as tentacles extending southward from the vast drainage basin of the Columbia River, the colossus of the Northwest. The fluid tentacles seemed to claw at the low northern rim of the basin, pushing it southward in a gently arcing retreat into Nevada.

Only along the southern rim is the boundary ragged and indistinct. Here the deep incision of the Colorado River slices westerly from the high plateaus and gnaws deeply into the edge of the basin. The master drainage of the American Southwest has deeply penetrated to the north, connecting some of the valleys of southwestern Utah and southeastern Nevada with the ocean. The deeper the Colorado cut through the shales and sands of the plateaus, the steeper is the course of its tributaries. This high gradient gives power to the streams, and they have sliced deep canyons into the southern heartland of the Great Basin.

It was only later that I realized my youthful picture of the Great Basin was completely incorrect. The boundaries were accurate, but a featureless, sand-filled bowl with lofty mountainous rims it is not. On the contrary, the floor of the bowl is broken into high snow-capped mountain ranges separated by broad elongated valleys. Although the drainages never reach the sea, it is by no means a lowland. The valley floors are generally between 4,000 and 5,000 feet above sea level, and the intervening mountains tower up to 13,000 feet. The valley floors in the central part of the basin are higher than along its flanks. The Great Basin resembles an upside-down, broken, and cracked bowl. No surface water leaves the Great Basin except by evaporation. Biologists and anthropologists have confused the usage of the term "Great Basin" by utilizing other criteria to define Great

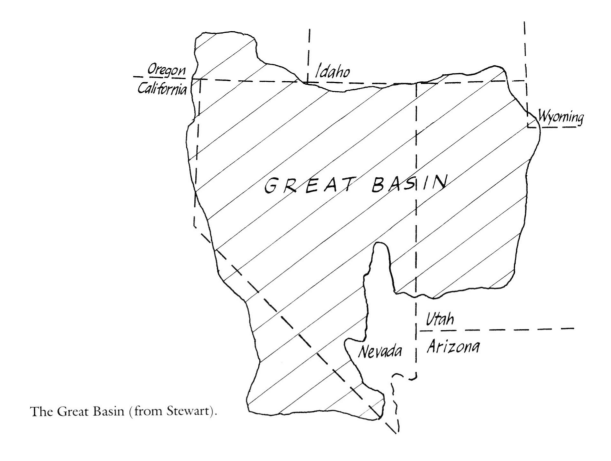

The Great Basin (from Stewart).

Basin biota or cultures, but the definition of the geographic region is founded only on hydrology.

Geologists have also inadvertently added to the confusion involving the definition of the Great Basin. We recognize the geographic and hydrologic character of the region. However, we include it as the northern portion of a much larger geologic province with a name so similar as to often cause perplexity even among its residents: the Basin and Range. This geologic territory includes virtually all the Great Basin and extends south and east through Arizona, New Mexico, and Texas all the way into Mexico. The Basin and Range can be differentiated from its neighboring geologic regions by its uplifted and tilted ranges separated by broad elongated basins. Southern New Mexico and Arizona have the same geologic style as Nevada.

The hydrologic Great Basin, then, is that portion of the geologic Basin and Range with no drainage to the sea. The two great rivers of the West, the Colorado and Columbia, are gnawing away at the flanks. But the rainfall is too sparse, the rivers have too little energy, and the uplift of the inverted bowl is too recent for the big rivers to breach the defenses of the Great Basin—yet.

In fact, the boundaries of the geologic Basin and Range do not exactly coincide with the hydrologic Great Basin. There are slight variations along all the borders except the southern edge. Here the Great Basin and Basin and Range boundaries are sharply

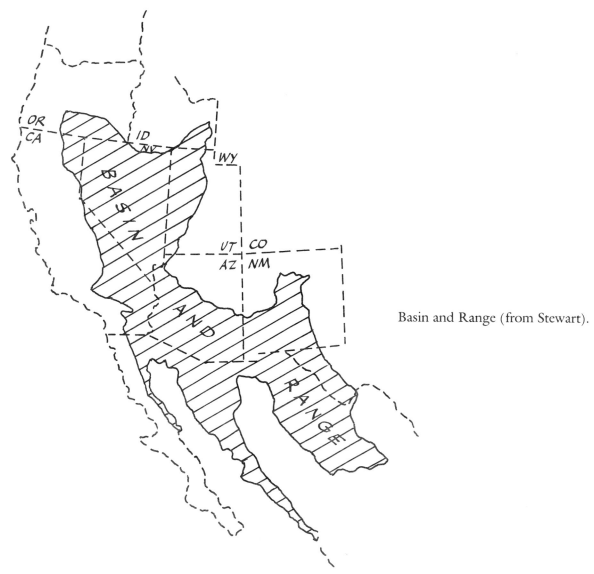

Basin and Range (from Stewart).

divergent. The Colorado River has managed to slice across the trend of the Basin and Range, and the northward-reaching tributaries deeply incise the southern boundary of the Great Basin.

The neighboring geologic provinces are sharply contrasted to the tilted and topsy-turvy basin-and-range geology. To the north, young lavas of the Snake River Plain buried the Basin and Range beneath thousands of feet of almost horizontal volcanic effusives. These black flows poured southward into northern Nevada.

The Colorado Plateau bounds the Basin and Range to the east. This is a geologically stable region of brightly hued rocks that have recently been uplifted vertically almost one mile with virtually no deformation of the horizontal layers. Rivers have deeply incised the uplifted Colorado Plateau into some of the most scenic canyons in the United States. Many have been designated as national parks. This is the domain of Grand Canyon, Bryce Canyon, and Zion. To the south, the Basin and Range merges into the mountainous structures of central Mexico. To the west of the Basin and Range lies the Sierra Nevada geologic province. This province is characterized by massive intrusions of once-molten rock that has cooled into the tough, erosion-resistant core of the Sierras.

3 Raw Materials of the Great Basin

I HAVE NOT ALWAYS BEEN a geologist. In fact, I had every intention of being a biologist. I have always been fascinated, even as a child, by the variety and adaptations of life around me. Biology is the science of excitement. Life abounds with freshness and the prospects of discovery. I had a high school friend who was destined to be a geologist and is one today. How boring were his interests. He collected dusty stones that sat in dignified rows on musty shelves in his basement, each rock identified by a jaw-breaking unpronounceable name. How static, how dull, how sedentary. I housed living snakes and young birds in my pockets and bedroom, to my sister's dismay, and I viewed scudgy green slime through my microscope. I learned bird calls and frog croaks. I identified heaps of wildflowers and ferns and then studied them in the bogs or shaded woods to discover when they bloomed and how they survived. How exciting was life, and how sharp the contrast with boring stay-at-home rocks. How profound was my ignorance!

Several years later, under the skillful guidance of a stimulating teacher, I discovered how much I had missed. The rocks held mysteries. They spoke of ancient times, of great floods, deep oceans, crashing surf, strange life forms, and groaning glaciers. I just didn't know how to ask the right questions or decipher the strange language. I was blind to the clues.

As I learned to unravel the secrets stored in stone, I realized that here were the real stories of earth. All the great dynamic past, all life, including giant snakes and ferns as large as trees, dinosaurs and monster sharks, enormous frogs and croaking scaled birds, all the natural world through all time was locked up in rock. Rather than explore merely the life of today, I could expand my lifetime into the past. I could discover hundreds of millions, even billions, of years of life and earth change. I wouldn't live just four score and ten. My experience would span billions of years.

A geologist was born, with the deep appreciation that rocks were story books. We just had to learn to read and to listen. As the beauty of the *Iliad* lies locked away from one who can't read Greek, so is the beauty of the earth's past hidden from those who can't read rocks.

MINERALS

Minerals are the stuff of the earth. They are the raw combinations of chemicals that formed during the cosmic origin of our planet. More than three-quarters of the matter of the earth is composed of only two elements: oxygen and silicon. Simplicity is considered by many to be the underlying theme of nature, but variety of form often evolves from the simplest materials.

More than two thousand minerals have been named and described by geologists, but even this seeming complexity is underlain again by simplicity. Fewer than a dozen minerals make up most of what we see and walk on every day. Some minerals are geologic jawbreakers, such as zvyagintsevite, picropharmacolite, or symplesite. Others are commonplace, as quartz, feldspar, and mica. In their pure forms, they are collected by museums and rockhounds. Those that are the hardest, or the most crystalline, or of the purest color are known as gems of the earth, like diamonds, rubies, sapphires, and emeralds. Some minerals such as gold, silver, and platinum are soft but are of sufficient rarity to be treasured by our culture. Others such as copper, lead, zinc, and molybdenum are required in our industries. We mine the available economic concentrations. Precious gems and metallic minerals, earth

Chalcanthite. Crystals from the Majuba Hill mine in northwest Nevada. These copper sulfate growths are about 1 cm. long. *Eric Offerman.*

Southeast Oregon lava flow. One of the major rock types in the Great Basin, igneous or once-molten rocks cover most of the northern borders of the region. *Rick Stetter.*

materials deemed valuable by our species, are formed in rocks that have undergone a variety of experiences. The Great Basin has an exciting and varied history. Great uplifts, collisions of continental masses, fiery volcanos, and burial beneath oceans have all created or altered the minerals of this region. Here the rocks are greatly experienced and here there is variety and mineral wealth in abundance.

ROCKS

If the minerals are the stuff of the earth, then rocks are the bricks made from it. Minerals combine to make the building blocks that form the mountains; they compose rocks. Some minerals have affinities with one another. They are conformists traveling in cohesive groups, all with the same chemical craving and all products of the same environments. Take a magma, a mixture of molten chemicals, and slowly cool the liquid. Crystals form as chemicals combine. The first to congeal are those that can no longer remain liquid as the mass cools. They crystallize and solidify into a mineral—a compound, often in crystal form, with set chemical characteristics and therefore prescribed physical characteristics. As the magma cools, other minerals with slightly different temperatures of crystallization form as a mixture with the first. If, suddenly, the magma solidifies before all the crystals have formed, the visible crystals will be dominantly those siblings that cooled first.

Some minerals form within restricted areas along the margins of the ocean, perhaps in a gulf. As the sea water evaporates, the dissolved chemicals concentrate. Those that precipitate are precisely in equilibrium with the drying conditions. The chemical controls are exactly the same as those within the magma. Only the temperature is different. This cool mixture of minerals must form an oceanic rock. You would not expect to find, nor would you, the formation of a magmatic rock in the swash of the gulf, or an oceanic mineral in the volcano's cauldron.

The lithologic Sherlock, seeking the environment in which a rock was formed, studies the minerals and their associations and, though elementary, the answers are often not obvious. Fossils often aid in deciphering the original environment. Some clams can only survive in salt water. Some species of snails live only on land. However, much of the geologic history of the Great Basin precedes the advent of life. Many of the more recent rocks formed at temperatures and pressures inimical to life. There are no fossil clues preserved in igneous or metamorphic rocks that suggest the environment at the time. But the minerals in all rocks contain chemical stories—tales of earlier environments and time. Any of us, with some persistence and basic deduction, can unravel these ancient tales.

Years of practice in sleuthing out rock stories by geologists have yielded the geologic stories of the Great Basin. From such discoveries geologists have also ferreted out the raw minerals and petroleum that fuel our economy. But on a more personal level, the investigation process also has created geologists who are excellent writers of whodunits or speculators on the stock market. Taking random fragments of data and piecing together a collage with most of the pictures missing can yield interesting and profitable results.

GREAT BASIN ROCK TYPES

Let's unravel the mystery. Geologists define three general environments in which rocks were formed, and the rocks of each are reasonably characteristic. They are either the product of cooling molten material, or they are the reconstituted remains of former rocks, or they are the recrystallized remains of preexisting rocks that have been either baked or pressurized. The rocks found in the Great Basin have all formed in one of these general environments. From such simplicity arises variety.

Fiery Origin Rocks

In the early feverish days of earth history there was but one environment of rock formation: cooling from a molten state. All the newly hardened rocks were igneous rocks. The cauldron fires still burn only a few tens of miles below the solid earth on which we tread. The earth is a plastic ball skinned over with a thin, brittle, solidified crust. As the earth

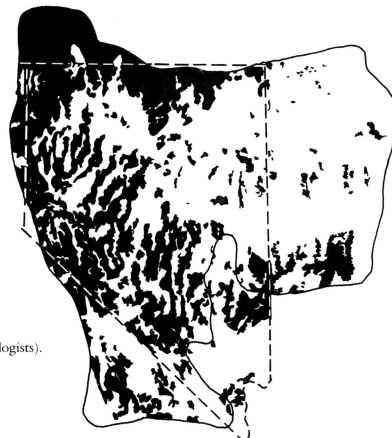

Igneous rocks of the Great Basin
(from American Association of Petroleum Geologists).

cools, the interior seethes with convective motion. The plastic internal materials flow upward and crack the brittle crust. The fractures in the broken crust serve as conduits along which molten materials ooze upward from the depths and intrude the crust.

When molten materials intrude toward the surface, they are faced with a fundamental choice: they either cool within the earth or on its surface. If they cool internally, they are surrounded by insulating rocks, and they cool very slowly, perhaps over hundreds of thousands of years. Slow cooling allows the crystallizing minerals time to form in chemical and crystalline purity. The slow accretion of chemicals to the crystal lattice allows large crystals to grow.

I remember as a child growing rock sugar candy. My sister and I would fill our jars full of saturated sugar water, hang a string in the solution, and put them away in our closets to wait. Like many a young lad, I couldn't contain myself for more than a few days. I would pull out the string and chew off the small, newly formed crystals. But my sister had incredible patience. Days or weeks later, she would pull out her string from the mixture. Her patience allowed the creation of cheek-bulging crystals, much to my chagrin. Intrusive rocks also have great patience. The insulation of the earth allows them to grow characteristically large and often perfectly formed crystals.

Promethean excesses sometimes force upward-moving magma to leak or blast out onto the surface. The frozen residues of these molten fragments are also igneous rocks. Although chemically the same as their intrusive kin, these volcanic rocks are markedly different in appearance from their more reticent interior cousins. Mercurial changes in environment stamp the rocks with profound distinction. Eruptive excess is often followed by sudden solidification. The molten mass is quickly chilled by the cold

Death Valley. Sediments lie scattered in rubble below yellow hills of soft sedimentary rocks. The materials of the earth continuously change form through geologic time to adapt to new conditions. *John Running*.

Death Valley. Metamorphic rocks represent change. Rocks once sedimentary and layered beneath an ocean are now altered by pressure and heat within the earth to form new rocks, metamorphosed into equilibrium with a new environment. *John Running.*

	TEXTURE	ROCK TYPES		
		High Silica Low Density Light Colored	Intermediate	Low Silica High Density Dark Colored
INTRUSIVE (deeply buried bodies)	Coarse-grained	Granite	Diorite	Gabbro
EXTRUSIVE (volcanic or shallow intrusive rocks)	Fine-grained	Rhyolite	Andesite	Basalt
	Glassy	Obsidian Pumice Perlite		
	Fragmental	Tuff Volcanic breccia Agglomerate		

Classification of Igneous Rocks.

rock surface over which it flows. The minerals freeze before they have an opportunity to grow into large crystals. The resulting rock is dense or even glassy, with few or no visible crystals. Lava flows and volcanic ash are made up of such materials. These extrusive igneous rocks are the quick-frozen result of lithologic impatience.

Volcanic violence has punctuated our geologic past in the Great Basin. Molten excesses have pierced the crust. Intrusive and extrusive igneous rocks are the norm in our mountain ranges.

Fortunately, there are certain similar characteristics between the intrusive and extrusive igneous rocks. Since they share the same fundamental chemistry, their classification systems can be shared. The rocks are named on the basis of their chemistry, but the nonchemist can still sort out most of them. Some have an abundance of lightweight silica. This creates a low density rock with only a few dark iron- or magnesium-bearing minerals. Others, silica-poor, are dense and dark. They are rich in dark ferromagnesium minerals. The names reflect this simple difference in chemical components. But all is not black and white in the real world of rocks. There are many shades of gray, and we have names for the gray rocks as well. If you would like to be on a first-name basis with a few once-hot rocks, you might try introducing yourself to the rocks in the igneous classification table.

Recycled Rocks

Sedimentary rocks are our neighbors, formed in environments that surround us. Consequently, with a little extra effort in observing our neighborhood, we can easily learn their names. More important, we can usually decipher the secrets of their past.

Sedimentary rocks are secondary rocks. They are derived. They result from the destruction of pre-existing rocks. There are two primary erosive processes that destroy old rocks and ready them for recycling into new rocks. These processes, mechanical demolition or chemical breakdown, are defined on the basis of the tools used in the attack.

The mechanical process breaks down old rocks

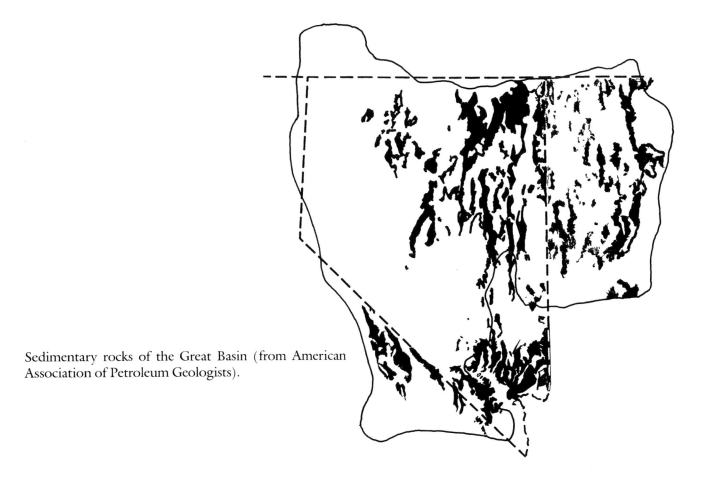

Sedimentary rocks of the Great Basin (from American Association of Petroleum Geologists).

into fragments. Cement these fragments together again and you have a new rock—a clastic sedimentary rock. An example is conglomerate: pebbles frozen into rock.

Clastic rocks are named according to the size of the fragments. The naming process is indifferent to the composition of the pieces, although any remnant information about the earlier rock types often gives additional clues about the environment. A pile of green peas cemented together would qualify as a clastic sedimentary rock, although surely one of unusual origin.

Each different environment leaves a different calling card within the rocks. For instance, some environments have great energy and tumult. The surf zone of the sea or a river in spring flood will wash out all the fine sediments and only large cobbles or boulders will resist the force of the water. These large fragments are in equilibrium. They remain as a

boulder bed after the winter storm or flood waters pass. A layer of conglomerate results. Environments of quiet repose, represented by deep ocean currents or slow-moving rivers, will leave behind fine detritus that settles and remains. The clastic rocks, therefore, reflect levels of energy.

Clastic sedimentary rocks, since they form at the earth's surface and are continuously forming today, are found virtually everywhere in the Great Basin. Whole mountain ranges of ancient clastics have been uplifted by earth movements. The shales, sandstones, and conglomerates lie like textbooks with their pages open. History is waiting to be deciphered.

Chemical breakdown is the other great erosive tool. Some rock materials are soluble. When they are out of chemical equilibrium with surrounding fluids, they dissolve. Atom by atom they disappear into the solution. They move within the fluid. Sometimes it is a trip of a few inches, or perhaps the

Quartzite. A rock, ground to a thin section less than the
thickness of a piece of paper, reveals its internal structure
under a microscope. X18. *K. B. Ketner, USGS.*

CLASTIC		
PARTICLE SIZE	SEDIMENT	SEDIMENTARY ROCK
More than 2mm	Boulder, cobble, pebble	Conglomerate
2mm – $\frac{1}{16}$mm	Sand	Sandstone
$\frac{1}{16}$mm – $\frac{1}{126}$mm	Silt	Siltstone
Less than $\frac{1}{126}$mm	Clay	Shale or mudstone

NONCLASTIC		
SILICEOUS (scratches glass)	LIMEY (fizzes in dilute acid)	OTHER
Chert Flint Chalcedony	Nonfossiliferous limestone Tufa (spring deposit) Travertine	Halite (salt) Gypsum (can scratch with a fingernail)

ORGANIC
Fossiliferous limestone Coquina (fossil clam) Radiolarian (silica microfossils) chert Peat Coal

Classification of Sedimentary Rocks.

atomic journey may continue for thousands of miles. Some may reach the ultimate sump—the ocean.

The laws of physics require that what goes up must come down. The laws of chemistry require that what dissolves must eventually precipitate. Somewhere on the journey, or perhaps on the floor of the sea, the atoms combine and drop out of solution. New rocks are thus formed. These are also sedimentary rocks—the recombination of the constituents of earlier rocks. These are the nonclastic sedimentary rocks.

It may not be obvious but igneous rocks and nonclastic sedimentary rocks have much in common. Both rock types result from crystallization from a liquid. One might precipitate at temperatures of thousands of degrees under the crushing pressures

of miles of rock. The other might precipitate from the dripping water faucet in your kitchen. The minerals and their physical characteristics usually reflect this difference.

Conditions of precipitation for different minerals are unique. Salt will not associate directly with gypsum. Differences that may appear subtle to us are vital to the minerals. Slight changes in temperature, pressure, or concentration will determine the orderly succession of mineral precipitation. Such discrimination usually results in the deposition of rocks composed of one mineral.

I have vivid memories etched indelibly into my consciousness of experiences with Great Basin nonclastic rocks. So did the pioneers. The shimmering white beds of salt, stretching to the horizon in the

FOLIATED		NON-FOLIATED	
Width of Folia	Rock Type	Chemical Composition	Rock Type
Wide	Gneiss	Siliceous	Quartzite
Medium (about like a sheet of paper)	Schist	Limey	Marble
Narrow	Slate		

Classification of Metamorphic Rocks.

snow-white Humboldt Sink, are nonclastic sedimentary rocks. So are the vast salt flats of the Devil's Golfcourse and Badwater in the interior of Death Valley. In both areas, mobile atoms of salt are stopped on their course to the sea. In the interiors of Great Basin troughs, where no rivers lead out to the ocean, the soluble constituents in groundwater and surface streams precipitate. They are stored in beds, sometimes thousands of feet thick, patiently waiting for the rupturing of the boundaries of the Great Basin. Someday, when rivers have carved their way through barrier mountain ranges or when the sea floods onto the continent, the patient crystals of salt will dissolve. Nonclastic sedimentary rocks are never at permanent rest until they reach the oceans. And even there, the illusion of permanence will be disrupted by uplift or oceanic transgression over the land. Every rock—all matter—will recycle.

Organic sedimentary rocks are composed of the fragments of plants or animals. Fossils or woody fragments lie encased within the surrounding rocks. This matter, also, must recycle.

CHANGED ROCKS

All metamorphic rocks are in disguise. They were once another rock type, either an igneous, sedimentary, or perhaps an earlier metamorphic rock. Their change did not require physical or chemical destruction. Their metamorphosis is not the result of erosion, but of recrystallization.

Existing igneous, sedimentary, or perhaps even metamorphic rocks might be blanketed by younger sediments. Deep burial increases heat. Or the rocks may be forced downward deep into the hot crust by the heavings of lithic stress. Perhaps continents collide, and rocks at the site of impact are crushed by great pressure. Or rocks may infortuitously come into contact with masses of rising intrusive magma. Metamorphism results.

Heating and pressuring cause recrystallization to a new set of minerals that are accustomed to the elevated temperatures and immense pressures. Metamorphic rocks are at home in such environments.

Rocks usually have a variety of minerals. The high heat and pressure of metamorphic environments frequently align these different minerals into parallel layers, or folia. Random orientation becomes organized. These folia allow geologists to conveniently segregate metamorphic rocks into two varieties: foliated and nonfoliated. We further divide the foliated metamorphics by the width of the foliation bands.

Not all rocks have mineralogic diversity. Some, like the nonclastic sedimentary rocks, precipitated from an original aqueous fluid of simple chemistry. Or a great stability in the environment of deposition may be reflected in the original clastic or igneous rocks by a simple chemistry. The ultimate in simplicity results in a monomineralic rock. Such a rock undergoing metamorphism wouldn't have the chemical and mineralogical complexity to differentiate

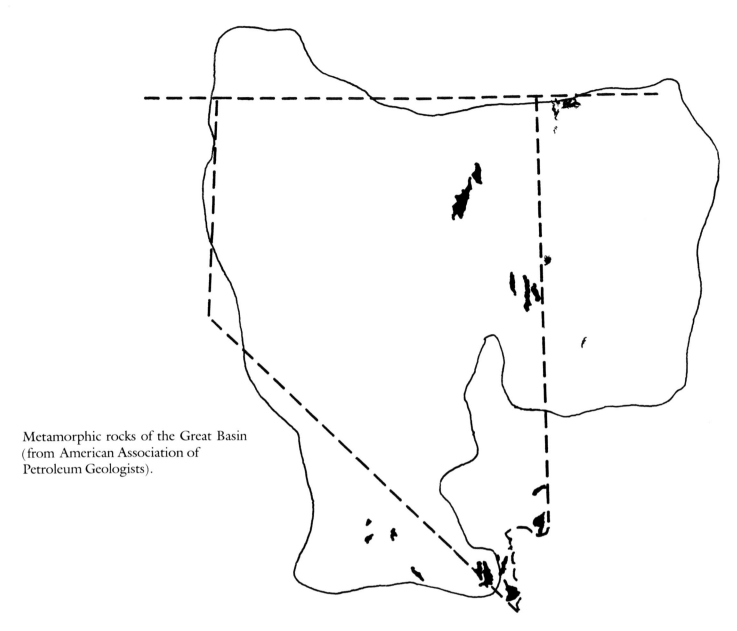

Metamorphic rocks of the Great Basin (from American Association of Petroleum Geologists).

into layers through the pressuring action. Nonfoliated metamorphic rocks result. Limestone, a monomineralic rock, will metamorphose into nonfoliated monomineralic marble. Limestone and marble look remarkably similar, as you might expect. The larger recrystallized crystals in the marble often allow you to make a distinction. Otherwise, the easiest way to tell a metamorphosed monomineralic rock from its unmetamorphosed progenitor is to check out the neighborhood. If the suspect rock is surrounded by obvious metamorphics, then you might assume guilt by association.

Metamorphic rocks are the most unusual of the three major rock types in the Great Basin. Only a few areas are dominantly metamorphic. If you want to study and observe metamorphics, you would be well advised to travel to a range where they predominate so the whole variety of metamorphic types can be studied.

These are the basic environments for rocks on earth. If you are interested in identifying a particular rock or mineral, there are excellent field guides available in local bookstores to help you determine the name of a particular specimen.

Valley of Fire. Most rocks are deposited horizontally. Sand piled high in dunes, however, is shifted and moved by the wind. This often produces inclined layers. *John Running*.

4 Understanding the Earth

UNDERSTANDING THE EARTH seems to be far too big an order for a mere mortal to attempt. The presumption that we might actually be able to comprehend the materials and the functions of the earth smacks of an egotism that only humans could suggest. Faced with the complexity of design in just one mountain range and the seemingly infinite variety of minerals and rocks in a single outcrop, it is incomprehensible to most people that we might actually be able to decipher the mysteries of even that part of the earth that we can see. Most of the earth's surface is buried under the seas. That which is exposed is only the smallest portion of the total, and we have only been able to penetrate a few miles with shafts or drills. The layperson could be tempted to abandon the quest before making the first groping attempts. Surely it requires an understanding of technical terminology and concepts far beyond the grasp of the nongeologist.

But no. The fundamental concepts of science are simple. They often seem to be hidden by scientists from the public, perhaps because the removal of the veil destroys the mystique. When the layperson discovers how simple science is, it may perhaps damage his or her faith in its omniscience.

Geologic concepts are simple. They are commonsensical and intrinsically logical, although they have not always been considered so. Each was revolutionary when first proposed and was in direct confrontation with a widely believed dogma. Years of argument and debate preceded their general acceptance by scientists. Today, they are so widely understood and accepted that there is seldom any question of their validity.

More than six hundred years ago, William of Occam (or Ockham), a British theologian, expressed a basic scientific principle after a lifetime of observing

natural processes. Occam's Razor is today considered by many scientists as an underlying concept of all science. His hypothesis states very plainly that the answers to most questions are fundamentally simple. Given a smorgasbord of possible explanations, pick the simplest and you will least likely suffer intellectual indigestion. Occam's Razor may be a powerful tool in the understanding of the way things work. You need not utilize spacemen or extraordinary parapsychology to explain events that can be easily explained in a simple way. My students say that Occam's Razor cuts through the baloney. Thoreau, in his laconic manner, said it best: "Simplify, simplify, simplify."

How do we go about solving the mysteries of the earth? Common sense and simplicity are the keys. A curious person who develops an ability to see behind the obvious will intuitively understand the drama of geology and may even consider it obvious. The ideas, once released from scientific jargon, are beautifully simple and concise. The verification of these concepts and explanation of the exceptions required years of sandy bedrolls.

A simple observation is that most sediments are deposited in horizontal layers, or beds, that are almost flat. Loose material simply cannot pile up with high slope angles because gravity constantly moves the particles downslope. Muds settling to the bottom of the ocean or a lake will usually lie along the bottom in a roughly horizontal fashion. This is particularly true for sedimentary rocks, which have settled through either air or water. There are many exceptions to this rule, as when lava flows down a steep hill or magma intrudes pre-existing rocks. But since 75 percent of the earth's surface is composed of sedimentary rock, the odds are certainly in your favor if you presume that a bed of rock was originally horizontal. Consequently, most layered rocks were once nearly flat. This concept of original horizontality allows us to interpret earth movements in areas where layered rocks are contorted or broken. Any tilted bedded rocks can be presumed to have moved since they were originally laid down. Conversely, most tilted and faulted rocks have shifted since they were deposited.

The lowest layer in a sequence of layered rocks is the oldest. How logical; the foundation must be built before the walls or roof, so the rocks first deposited will underlie those that are later deposited. As long as the rock materials were deposited at the earth's surface and were not disturbed by later crustal movements, this concept is rarely violated. Observing a cliff with stacked layers of rocks, you can assume that the oldest rock layers will usually be found at the bottom of the cliff. If you observe a sequence of tilted rocks, walk in the direction of the tilt. You are walking through time. As you move, you are progressively moving through younger rocks. This superposition of the age of layered rocks allows us to decipher the sequence of past events in proper order.

The fossils in the rocks will also become younger as you go higher in the layered sequence. Evolutionary change is clearly printed in the fossil remains. The concept of faunal succession is a corollary to superposition.

Another concept of geology is described as uniformitarianism. This pretentious word, which sounds like a religion, merely states that the present is the key to the past. It is this concept that allows everyone who is observant to think and speculate as a geologist. In the middle of the eighteenth century there was a raging debate in the scientific and religious communities regarding this concept. A Scotsman, James Hutton, was in the center of this dispute. He stated that the processes that we see acting on the earth around us today are the same processes that were active in the geologic past and they will be the same processes of the future. Hutton envisioned an immense length of geologic time to account for the mountains, valleys, and physical features of the earth. The earth was in a general state of continuous change. The changes were occurring so slowly they were rarely observed within the lifetime of a human or the lifetimes of generations of humans. This concept was diametrically opposed by the dominant religious views of the western world at that time. Archbishop James Ussher had calculated during the seventeenth century that the earth was only six thousand years old. He did this by counting up the generations of the Bible since Adam. Later theologians were able to refine his date to 9:00 A.M., October 26, 4004 B.C. Similarly, religious views expressed the concept of catastrophism as the primary modifying effect on the earth. Noah's flood, the great worldwide catastrophic event that accounted for the presence of fossils in rocks, was a recent phenomenon only a few thousands of years ago. Hutton's ideas of uniformity of process and enormity of geologic time were thought to represent a direct affront to religious dogma. He wrote of "no vestige of a beginning, no prospect of an end." This long-range view of general uniformity of process allows geologists to sleuth out the past.

Rocks, like people, are the direct result of both their internal structure and the environment in which they form. They preserve the evidence of their past experience as clearly as does the corporate executive or the geologist. Rocks reflect their environment. The salt beds rimming the Great Salt Lake are different from the gravels and sands on the shores of Lake Mead. Should the present sediments be preserved as rock, the rock salt and the sandstone would contain the characteristics of the different environments. Fortunately, many rocks form in the familiar environments around us—beaches, lakes, streams. If we are careful observers of our neighborhoods, we can decipher stories of our neighbors' past. The rocks around us have familiar stories to tell if we take the time to ask the right questions and have the confidence that we will understand the answers. The idea of uniformitarianism allows us to use our experience to extrapolate that of the rocks. To speak of what today's processes are is to speak of what they were.

Death Valley. Streams form deltas on the valley floor. This process has been active for billions of years, according to the geologic record. Billions of years from now, streams will continue to build deltas. Uniformity of process underlies geologic concepts. *John Running*.

EONS	ERAS	PERIODS	EPOCHS	Years Ago (in millions)
Phanerozoic	Cenozoic	Quaternary	Holocene	
			Pleistocene	
				3
		Tertiary	Pliocene	7.5
			Miocene	26
			Oligocene	37
			Eocene	54
			Paleocene	
				67
	Mesozoic	Cretaceous		
				130
		Jurassic		
				200
		Triassic		
				237
	Paleozoic	Permian		
				293
		Pennsylvanian Carboniferous		320
		Mississippian Carboniferous		356
		Devonian		400
		Silurian		431
		Ordovician		
				495
		Cambrian		
				586
Cryptozoic	Prepaleozoic			
				3,500
Azoic				
				4,500

The Geologic Time Scale.

5 Time, the Relentless Flow

THE MOST significant contribution of geology to human knowledge and understanding of the earth is our concept of time. No single scientific concept has so altered our view of geologic processes or expanded the potential of possibilities than the availability of vast oceans of time. Given sufficient time, virtually anything is possible. Mountain ranges rise and are pulverized into fragments. Oceans fill and drain like backyard puddles in rain squalls. Glaciers bury continents under miles of ice and then disappear. It takes time to work geologic processes, but this is one commodity the earth has in abundance.

There are many ways to understand time. One way is to relate events to other known occurrences. For example, the intrusive, once-molten rocks of the Sierras cut through other beds. They must be younger, since the surrounding beds had to be there first to be intruded. Horizontal rocks high in the Snake Range are usually younger than the rocks beneath them.

This is the world of relative time. One event occurs before or after another. Our geologic calendar has been carefully pieced together by scientists using relative ages of events. The names of the geologic eons and eras are based on the absence or presence of life in the rocks. Further refinements can be made based on the complexity of the life forms present. The Azoic-aged rocks are those without life. The Cryptozoic-aged rocks, as you might surmise, are those that contain hidden life, usually microscopic. The Phanerozoic-aged rocks are characterized by visible life. Those rocks with readily visible life can be broken down into three major units based on the relative stage of evolution of the life forms. Paleozoic rocks contain old life fossils. Mesozoic rocks contain middle life forms. The Cenozoic is the time of recent life.

The eras are divided into smaller time units. Most of the period names of the Paleozoic and Mesozoic are derived from geographic locations. The names roll off a geologist's tongue like a litany in church: Cambrian, Ordovician, Silurian, Devonian, Mississippian, Pennsylvanian, and Permian. The time units were named from regions where the rocks are especially prevalent. Triassic, Jurassic, and Cretaceous—the nomenclature of the Mesozoic. The Cenozoic periods are vestigial remnants of an earlier geologic attempt to differentiate rocks on the basis of the degree of hardness of the rock. Only the last portion of the previous venture remains—Tertiary and Quaternary. The Primary and Secondary names have been relegated to the geologic trash heap. Cenozoic rocks are especially abundant. They are the most recently deposited rocks and have had the least opportunity for erosion. They all share the common suffix -*cene*, derived from the Greek word *kainos*, or recent. They also have interesting prefixes of Greek origin to distinguish their relative antiquity and to make them sound sufficiently scientific. Perhaps, in this case, the Greek is clearer than the English and easier to remember. The litany of the Cenozoic: Paleocene, or "ancient recent"; Eocene, or "earliest recent"; Oligocene, or "scant recent"; Miocene, or "less recent"; Pliocene, or "more recent"; and Pleistocene, or "most recent." It is useful to become familiar with these names since they will be used throughout this book. They are the language of the history of the earth.

The ages of relative time divide the territory. But the immensity of the subject has long baffled scientists. One of the founders of geology, James Hutton, in 1775 addressed the Royal Society of Edinburgh on the age of the earth. Using the concept of uniformity of process, he concluded "that it had required an indefinite space of time to have produced the land which now appears."

But how much time? How much older or younger is the rock of the Sierras than that of the Wasatch? These queries lead to the world of absolute time, the world of days, hours, and nanoseconds. Many events in human history can be accurately measured in absolute time because there were people to record and observe the incident. Mount St. Helens erupted at 8:32 A.M., May 18, 1980, as reported by the geologist who died a few moments later when he was engulfed by the blast.

How do we determine absolute time for those millennia before there were people to keep records? Luckily, rocks have built-in clocks that tick off the time since the minerals hardened from magma or congealed from fragments. The clocks are the radioactive decay of isotopes bound into the minerals. The ticking is as accurate as your quartz crystal wristwatch. Decay rates are known, and accurate laboratory measurements can thus pinpoint, in years, the dates of geologic events. Of course, as with your watch, there is a slight inaccuracy, but when you are dealing in billions, a few millions are of small consequence.

Geologists, using radioactive clocks, determined the date of the rocks and added the years to our calendar of relative time. Our studies reveal that the oldest earth rocks are a little more than four billion years old. From the faint ticking of the clocks in moon rocks and meteorites, we surmise that the earth formed about a half a billion years earlier.

Life on earth is only a little younger than the oldest rocks. Signs of primitive life date back somewhat more than 3.5 billion years. Forms that resemble today's bacteria are found in those ancient sediments. Throughout the next 2.5 billion years life was restricted to the oceans and perhaps a few shallow ponds on the continents. Algae and bacteria are sparsely represented until about 700 million years ago, when jellyfish and worms began to diversify the fossil record. Suddenly, about 570 million years ago, life forms with easily fossilized skeletal materials began to proliferate. This enormous span of time, from earth formation to the first clam shells of 570 million years ago, we call the Prepaleozoic.

A large, single continent probably existed in the latter part of the Prepaleozoic. Reacting to the surging tumult of interior heat, the continent fragmented into pieces. Thousands of miles of new shoreline, with a great diversity of environmental niches, were formed. Perhaps the explosion of shelled life forms was in response to all these new and exciting choices for homesites.

The time between the rapid evolutionary expansion of marine life, which marked the end of the Prepaleozoic, until about 230 million years ago we call the Paleozoic, or old life era. This was the time of trilobites, clams, and fish. Plants spread from the seaweed-strewn shores into the terrestrial world above the tides. Greenery quickly clothed the rocks. Soon thereafter, amphibians slithered out of the sea and colonized the land. They were the first of the evolutionary sequence of animals that eventually occupied almost every corner of the earth. Think of it. A land unoccupied by any animals, wide open, with a smorgasbord of good plant food ready for the taking. It wasn't long before every continent had its complement of critters, filling all the terrestrial niches. Then the catastrophe.

About 225 million years ago, the greatest tragedy to befall life on earth occurred. In a short span of only a few tens of millions of years, most major animal groups died off.

Why? We have no proven explanation, but we have several hypotheses. At that time, the disjunct wandering continents fused together into a single megacontinent, Pangea. The reduction in the number of independently moving plates on the earth's surface meant an absence of thermally swollen spreading centers along oceanic ridges. Without the swollen ridges in the midparts of oceans, there was more room in the ocean basins for sea water, and the seas retreated from the continents. Paleontologists suspect that the extinctions resulted from this sea withdrawal. The single continent also resulted in the unification and reduction of coastal and continental environments. There was a great competition

Calville Wash, southern Nevada. The angular junction of
beds is the result of earth movement and tilting of once
horizontal beds, followed by erosion and then deposition
of horizontal layers. Now the record is being destroyed by
erosion. *John Running.*

Pyramid Lake, Nevada. Patterns in nature are simple and repetitive. The veins of a leaf or a finger follow the same pattern as water eroding a soft mud. *John Running*.

for the few remaining good homesites. The most adaptable survived. Many species died off in the biologic struggle.

Another hypothesis proposes that the catastrophe resulted from drastic changes in the salinity of the ocean. Or perhaps it was due to a series of radical climatic changes. We do not know the cause. Whatever the reason, geologists use this catastrophe to mark the end of the old life era. This was the terminal point of the Paleozoic.

New characters now moved onto the earthly stage. Birds, dinosaurs, and a proliferation of incredibly ornate marine life characterize this middle life era, the Mesozoic. New groups gained ascendancy upon the graveyards of the old. But once again, global disaster occurred. Approximately 65 million years ago a second widespread extinction eradicated many species, including the last of the dinosaurs. What caused this disaster? There were many changes that alone or in combination could have caused it. The seas abruptly withdrew from the continents once again. Worldwide climates changed, and the earth became colder. New species of animals, the mammals, proliferated. Perhaps they fed on the vulnerable eggs or young of the older denizens. New plant species, with new strategies for survival, outcompeted the old and spread over the continents. Some have even speculated that the earth collided with an extraterrestrial body and the resulting chaos caused the extinctions. Perhaps all of these factors combined into a tragedy. Again, we don't know. This biotic disaster marks the end of another era. The Mesozoic is a time spanned by unexplained organic catastrophes.

The recent life era, or Cenozoic, takes us to the present and includes the debut of many mammal species. Grasses gained importance on the plains, and mammalian grazers diversified to happily munch the new resource. During the later part of this time, the great Ice Ages spread their freezing influence upon the earth.

Talking with geologists is like taking a trip through time. The events of millions or hundreds of millions of years ago roll off their lips as though they were discussing last week's birthday party. Beware of geologists' meaning when they rave in ecstasy about the beauties of a recent volcanic cone— it is probably only a few millions of years old.

How can we understand the enormity of geologic time? Perhaps relating geologic time to known measurements will help. If the 4.5 billion years of earth history were compressed into one week, with Monday as the first day, then the oldest rocks yet found on earth and the first life forms would appear almost simultaneously at Monday's twilight. So few ancient rocks exist that little else is known about the earth until the end of the week. The Paleozoic and its shelled animals began in the early dawn of Sunday morning. The old-life era continued until about 3:30 P.M., when the biologic disaster occurred. Then, in the gathering dusk of Sunday evening, the Mesozoic and its dinosaurs encompassed the earth. Suddenly, shortly before 10:00 P.M., the dinosaurs walked off the stage during the second extinction and the mammalian Cenozoic began. Ten minutes before midnight, at the very end of the week, the earliest people arrived. In the last fraction of a second, as the clock hammer falls to strike midnight, the Declaration of Independence was signed.

Another measure of time may help. Hold your arms wide apart from your sides. If the beginning of earth time is at your right fingertip, then life would begin near your right elbow. The Paleozoic clams would be at home from the middle of your left forearm to the beginning of your left index finger. The dinosaurs would cavort along your finger to the last joint. The end of your finger, from last joint to tip, would be the mammal years. And our species? We can be measured in the snip of a nailclipper.

The average person in the United States lives to an age of about seventy-five and one-half years. If we relate the age of the earth to our average lifespan, with the earliest time of earth equal to the birth of a child and the present corresponding to the end of the average human life, we have another analogy to perhaps better understand the antiquity of the earth. The earliest life forms we have yet found, primitive bacterial-like forms, would have

Fossil fish near Fallon, Nevada. Impressions left millions of years ago in soft mud startle and fascinate the finders today. Life is linked in an unbroken chain stretching back to the primeval ooze. *John Running*.

evolved about the middle of earth's eighth year. The kid was in third grade. The Paleozoic, when shelled life forms began, would have begun in early July of earth's sixty-fifth year, close to retirement for many of us. Earth was indeed mature when life finally evolved into more complex forms. The first footprints of a land animal would have sunk into the primordial muds in the early fall of the seventieth year. The earliest birds would have tried their wings about three years ago. The great dinosaurs reigned supreme from about the fall of three years ago until the beginning of last year, when they were snuffed out in a few moments of comparative time. The age of the mammals, the Cenozoic, would have begun during the last year of the life of earth. The recent Ice Age would have encompassed the last week and a half.

The earth around us gives the false impression of being in equilibrium. All geologic materials and forces tend toward a stability that is the exact counterbalance between opposing forces. This equilibrium is actually a dynamic balance, and the slightest alteration in the forces involved results in change—a new balance. We tend to view the scenery as changeless and permanent, but it is only the instantaneous representation of a continuum of change. It is the temporary balance of nature, and it

will evolve into new forms. The earth is not static. We live for such a short time relative to the span of geologic time that the dynamic world surrounding us seems permanent to us. A microbe, that might live for one-thousandth of a second, does not know you are alive if he crawls on your chest in the time between your heartbeats. We also have difficulty imagining the vitality of a world whose heartbeats are so widely spaced. If we blinked our eyes open for one second every thousand years what dynamic views we would have of our earth! Mountain ranges would convulsively thrust upward, aprons of gravel blanketing the mountain slopes would become conveyor belts of detritus, sea coasts would advance and retreat as ocean depths varied, new species would evolve and disappear—the dynamic interplay that we call the "balance of nature" would become vibrant and animated. To acquire our new eyes in viewing the landscape, we must change our notions of time. To see the changes in the earth reflected in our surroundings, we lengthen our life span through our imagination and visualize geologic events that occurred hundreds of millions of years ago. The evidence of these events is as clearly depicted in the rocks, and as easily imagined, as the evidence of the landing of the Pilgrims on Plymouth Rock or the California gold rush.

6 Fossils, the Timekeepers

IF THE HISTORY of the earth were recorded in a book, the pages would correspond to layers of rock. These layers would usually be sedimentary rock. If these geologic pages were followed sequentially, the history of the earth could be deciphered from the beginning to the present. Each rock layer, like the page of a book, is thin but covers a wide area. Rock materials, scattered by wind and water, generally do not pile up very high against the forces of gravity but spread in a thin sheet over the countryside. Geologists over a wide area read the same pages and interpret the same story. Using the concepts of superposition and lateral continuity, the relative ages can be worked out. The bottom layers of the geologic book would be the first to have been deposited. Later layers piled up on top.

Erosion, however, can destroy some pages from our geologic book. Missing layers are easily detected if earth movements accompany the erosion. The older beds will then be tilted and eroded before the newer layers are deposited horizontally over the warped oldtimers. Such an angular unconformity tells us that not only pages but perhaps whole chapters are missing. When erosion scours horizontal beds without attendant folding, the missing layers may be difficult to detect, since the horizontal younger layers lie in conformity with the older layers. The pagination of such a parallel unconformity is supplied by fossils. Geologists know the major fossil groups, and with additional training or experience they learn the index fossils of a particular time and thus can work out geologic chronology. Geologists usually work with groups or assemblages of fossils that would only have existed for a relatively short period of time.

The turning of the pages into the geologic past reveals life forms imprinted on virtually every sheet.

The oldest rocks on earth are only a little older than the oldest evidence of life. Paleontologists have dug and continue to patiently dig at the rocks to piece together the long record of life on earth.

Rambling through the ranges and rocks of the Great Basin, you will certainly discover some fossils in your travels. Virtually all the ranges that contain sedimentary rocks will have fossils. Knowing the age of the fossil can tell you the age of the rock. Originally, geologists only knew the relative age of fossils. We developed a time scale using fossils as guides. This relative scale was useful, but it was not easily related to either nonfossiliferous rocks or to the absolute age of the rocks. Now rocks containing fossils have been dated with radioactive decay methods so their absolute ages are known, and they can be correlated to nonfossiliferous igneous and metamorphic rocks.

Living forms are often severely limited in their range by environmental controls. A pismo clam will have trouble surviving in my backyard. Fossil remnants, therefore, may give us detailed information on the environments of the rock that encloses the fossils. Marine or nonmarine? Warm water or cold? Deep water or shallow? Equatorial or polar? Answers to such questions are often easily found. Looking for fossils is not only informative but fun. The detective game is played with once-living clues.

Preservation of fossils takes place in many ways. Soft parts are usually eaten by scavengers or predators, and those that survive the munching usually rot quickly. It is the indigestible hard parts, the bones, shells, and teeth, that are most often preserved.

Animals commonly use either calcium carbonate or silica for their shells or skeletons. Some, such as lobsters or crabs, use chitin for their shells. These materials, when discarded by the former owner, may slowly be converted into rock. This transformation is usually the result of replacement by chemicals circulated by sea water or groundwater. Chitinous shells are usually not replaced but they sometimes mark the encasing rocks with a carbon residue.

Fossils are useful to geologists in determining the age of the rocks and this allows us to correlate the rock layers not only within the Great Basin but throughout the continent. Some fossils are small enough to be brought to the surface in cuttings from the drills of oil wells, and they help geologists determine the ages of rocks from deep bore holes. When geologists can correlate rocks of a similar age from an entire continent, or the entire earth, it enables them to piece together the paleoenvironments of the ancient time. They can determine the geography of the distant past. Much of the geologic history of the Great Basin, such as water depth, distribution of seas and land, and climatic conditions, can be revealed by paleontologic study. Water temperatures can be discovered by careful analyses of isotopes, and even the length of days can be estimated by growth rings of fossil clams.

The rocks of the Great Basin reflect life forms that may date back more than two billion years. It is hopeless to even consider becoming acquainted with all the living beings that have lived here through that immense time. There are, however, some advantages in working out the genealogies of the more common or easily identified former residents. Reducing the number of fossils can simplify the numbers and allow you to recognize familiar faces in the rocks. To really decipher the complexities of evolution and environment, you need to become heavily involved in the field of paleontology. This can be a fascinating avocation.

During the first three-quarters of the time represented by the rocks of the Great Basin, very few species evolved. Those early residents were simple and soft. Their calling cards, represented by their earthly remains, simply did not fossilize well. There are also few remnants of these early inhabitants since their remains were swept away by erosion or were buried under thousands of feet of rock and metamorphosed. Thrust up a few mountains and explode a few volcanos, and it's easy to remove the evidence. Much of what was here is simply gone, baked, or hidden.

This leaves the residual few. Those species lucky enough to have been preserved are now exposed in the mountainsides or river-cut canyons of the Great Basin. Another idle flicker of geologic time, and their remnants would be destroyed forever. The few who survived the odds are more than enough to keep us busy for quite some time collecting, cataloging, and studying. Many a student has struggled through a thesis attempting to solve the puzzle, and more than one geologist has devoted a lifetime of study to the fragments left behind. Tens of thousands of species are identified, testifying to the arduous labor of paleontologists. Volumes of reference material have been written.

Faced with the complexity of Latin names and dusty bins of books, how can you sort out some of these past forms and be able to place the fossils you find into a named category? Another book could be written just to achieve this, but let's simplify the task by examining only the most abundant and easily recognized Great Basin fossils.

GREAT BASIN FOSSILS

Foraminifera

While carefully turning over a piece of gray limestone in your hands you may see some small fossils that look like miniature grains of wheat or rice. These are the one-celled animals called forams. They require a magnifying glass to really observe carefully, and they appear in many forms. Their shells are usually composed of limestone or silica.

The most common forams are called fusulinids. These are the ones that look like wheat grains. They are often so abundant in the limestone that the rock may be mostly composed of their remains. Fusulinids are very common in Great Basin limestone and are particularly characteristic of the late Paleozoic.

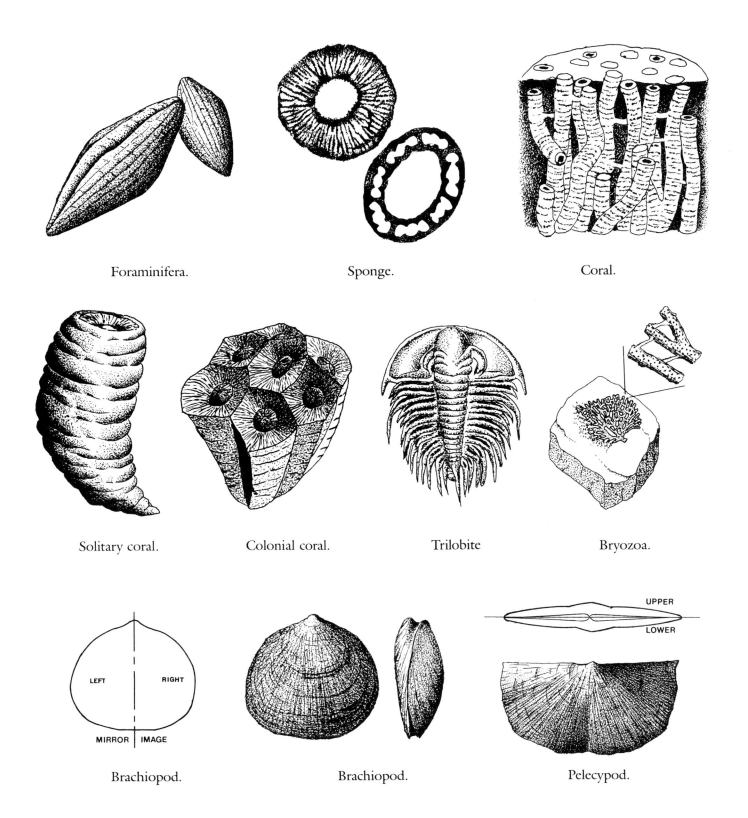

Foraminifera.

Sponge.

Coral.

Solitary coral.

Colonial coral.

Trilobite

Bryozoa.

LEFT RIGHT

MIRROR IMAGE

Brachiopod.

Brachiopod.

UPPER

LOWER

Pelecypod.

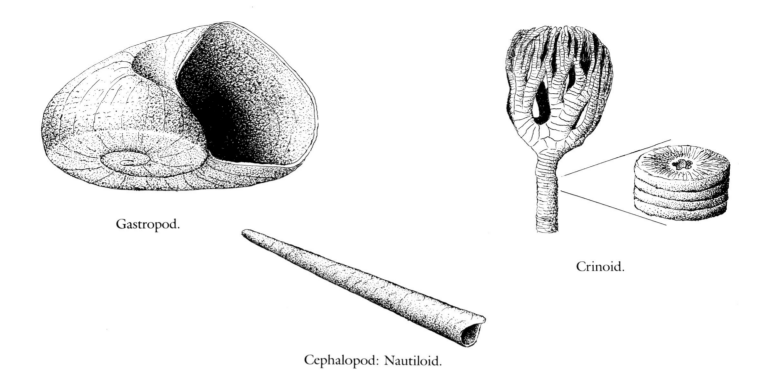

Gastropod.

Cephalopod: Nautiloid.

Crinoid.

had many snails and the lake muds are often littered with their shells.

Cephalopods

Cephalopods are highly advanced animals with excellent eyesight and nervous systems. These are the squids, octopuses, and cuttlefish that are abundant in today's oceans. They are relatively unusual as fossils in Great Basin rocks.

Calcium carbonate was used for the hard body parts. In the Paleozoic most cephalopods were straight-shelled. Although most were only a foot or so long, some giants grew as long as twenty feet. Coiling of the shells became common by the late Paleozoic.

Trilobites

These creatures looked much like some of today's crabs or the sow bugs in your garden. Both paleontologists and amateur fossil collectors find these to be the most appealing creatures for their collections. For a few dollars they can be purchased as keyfobs or earrings in national park tourist shops. Their great popularity may stem from the interesting segmented bodies that are often very well preserved in shales. Their bodies are divided into three lobate sections, hence the name "tri-lob-ite." They have been extinct for two hundred million years, and if the tourist shops have their way, their fossil remains are liable to be extinct in another twenty years.

The trilobites had excellent eyes, a jointed body, and a tail. Through the passage of time and evolution they became increasingly ornate. Their heads sported long horns and their tails grew into lunate spikes. Since they periodically shed their horny armor to accommodate growth, a single trilobite could provide many fossils.

During the Paleozoic, trilobites were among the dominant life forms in the oceans. The shales of the eastern and southern Great Basin, the former muds of the early Paleozoic oceanic shelf, are particularly rich with these fossils. Fine collecting has been done in the House Range, Utah, and near Pioche and Las Vegas, Nevada.

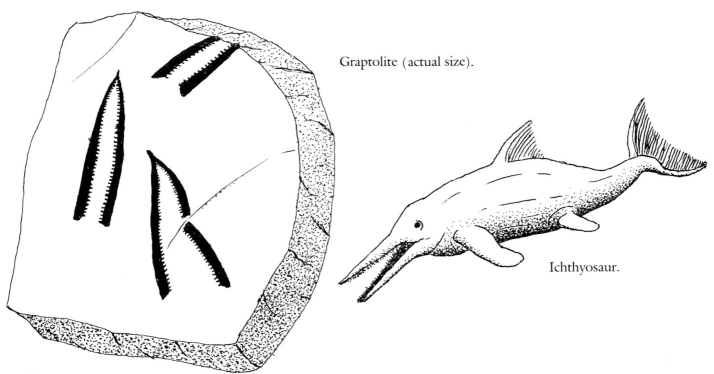

Graptolite (actual size).

Ichthyosaur.

Crinoids

These animals are often mistaken for marine plants, either sea lilies or sea palms. They do indeed look like palm trees rooted in the muds of the ocean bottom. Their stalked bodies, which resemble a tree trunk, rise to a crown of flailing arms that sift plankton from the sea. After the death of the animal, the disks that make up the stem fall apart and are carried around by currents and waves. These disks are common in the limestones of the Great Basin and look like small Cheerios.

Graptolites

These simple animals first appeared in the earliest Paleozoic, but they died out toward the end of the era. They were colonial creatures who lived in tiny cups located along slender stems. Some were attached to a round float or the stems were attached together. They floated worldwide throughout the Paleozoic oceans. In the Great Basin they are most common in the shales and siltstones of western Nevada. They look like small black lines with sawtoothed edges.

Plant Fossils

Petrified wood is common in many of the Mesozoic red beds of the Great Basin. Some wood is also found within Cenozoic ash falls, where silicic ash buried standing forests. Occasionally, leaf fossils have been found in Mesozoic or Cenozoic beds.

Vertebrate Fossils

The Cenozoic beds of the Great Basin contain rodent bones and teeth. The numerous small mammals have left traces in many of the basin-filling muds and gravels. Occasional fish bones or skeletons are discovered in lake sediments in the basins. More impressive vertebrate remains are also found in the region. The great swimming reptile, Ichthyosaurus, has a Nevada state park dedicated to the preservation of its fossilized remains. Mammoth and mastodon tusks and bones occasionally wash out of a river bank or gravel bed. Impressive finds have been unearthed in the Black Rock Desert in northern Nevada and in lake sediments near Las Vegas.

Red Rock Canyon, southern Nevada. Earth forces from plates in collision have forced the old dark limestones over the younger light-colored sandstones. *John Running*.

7 The Restless Earth

DURING the past twenty years, a new idea has swept the geologic profession. The concept is as revolutionary and has been as heatedly debated as the dispute over the age of the earth hundreds of years ago. The idea is termed plate tectonics. Behind the fancy name, the story is that the continents are adrift! And not just continents, but ocean basins as well. In fact, continents are only the emergent parts of large pieces or plates of the earth's surface. Both continents and ocean floors compose these lithospheric blocks.

Some plates are huge. The Pacific Plate extends from San Francisco almost to Tokyo and from Anchorage almost to New Zealand. Other plates are minuscule—so small that some geologists don't think they warrant the title. Tiny plates underlie the Fiji Islands, the Adriatic Sea, Iran, and Turkey. We even have, named in scientific seriousness, a China Plate. As in all fields there are splitters and lumpers. Some geologists delight in subdividing and naming ever-smaller plates. Others concentrate on the larger picture and define only a few. Most agree that the global surface can be divided into at least twenty discrete plates. Everything on the earth's surface is moving, and each plate is moving in a different direction from its neighbors.

Independence creates interesting contrast. Independence of plate motion gives the earth variety, creating the heights of the mountains and the depths of the oceans. Independence can also create instability and occasionally raise havoc, as when San Francisco crumbled into ruins in 1906 and Mount St. Helens self-destructed in 1980.

What moves these great slabs of lithosphere? The answer is debated, but most geologists believe it is the internal heat of the earth. Radioactive minerals decaying in the interior generate great heat and temperatures rise. The insulating effect of the overlying rock traps this energy, preventing its escape into the bitter cold of space. It is as though the earth is wearing a parka on a cold winter day. Eventually the rock becomes so hot that it can flow. The flowage is incredibly slow by human standards, since the deep magma under high pressure is viscous and sticky.

The heat, however, is unevenly distributed. Concentrated in some regions, molten rock swells upward from thermal expansion and raises great welts, called spreading centers, on the earth's surface. Rising plumes of hot magma crack the thin, stretched, upbowed, rigid plate overlying the spreading center. The pieces of the broken plate move apart. Each is tilted away from the other, like two trap doors opening upward. They slide off the thermal bulge on the slushy deeper magma. Such cracking causes shallow earthquakes, and hot magma erupts through the fractures as volcanos.

New crust is made at these rifts. Rocks form from the solidifying magmas and glue themselves to the spreading edges of the plates. These plates slide or are pushed downhill off the high spreading center, making room for more magma to heave to the surface. Crust, newly evolved and moving, is formed. These spreading centers make great chains of mountains, studded with volcanos and shaken by shallow earthquakes. The highest thermally swollen regions are rifted by the pulling apart of the plates. The longest continuous mountain range on earth is the result of magmas welling up along a spreading center. This range, the Mid-Atlantic Ridge, is visible in only a few places where the highest mountain peaks rise above the waters of the Atlantic.

The new crustal rocks, formed by volcanos at spreading centers, are dark and heavy. They are derived from the heavy iron-rich magma that sank

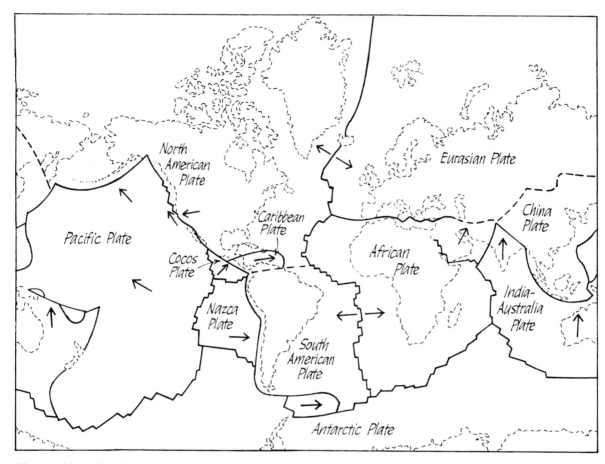

The earth's major plates.

into the depths of the earth billions of years ago. As the newcomers to the surface cool, they contract and become dense. Slowly they slide off the welt. They age outward from the ridge. With increasing age, they become cold and sink deep into the underlying slush. Rivers pour into these low regions to become the world's oceans. Ocean rocks, therefore, are new and heavy.

The oldest ocean floor we've drilled and sampled is only 150 million years old—the equivalent of two and a half years ago in the relation of a human life span to the earth. The continents, on the other hand, are ancient. The oldest rocks yet found on the continents are 3.8 billion years old. The earth was only old enough to be in elementary school when compared to human lives.

Continents are fundamentally different from ocean basins. They are lightweight, composed of low-density silica and aluminum. They float on rocks with a density similar to that of ocean floors, so they ride high, like icebergs, and they always have, since the earliest days of the earth.

When I was an undergraduate geology student, we were taught that ocean basins were permanent features of the earth—always had been, always would be. Now we've drilled them and dated the rocks in the soggy cores. Young. However, we know oceans have been on earth since the earliest days, since some of the oldest rocks on continents are clearly stamped marine. Where have all the old sea floors gone?

If plates move apart at spreading centers, and

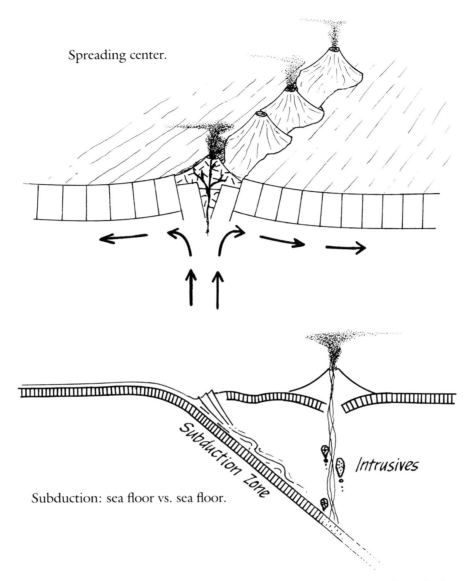

Spreading center.

Subduction: sea floor vs. sea floor.

Subduction Zone

Intrusives

they do, then it is reasonable to surmise that somewhere plates are in collision, and they are. There are three possible collision types: sea floor against sea floor, sea floor against continent, and continent against continent.

If the plates in collision are both dense ocean floors, which will yield to the impact? The plate margin farthest from its spreading center will be the older, colder, and thus more dense plate, and it will be forced under the newer, lighter plate margin. Rubbing, jostling, resisting, the old heavy plate

slides down. The sea floor is dragged down with it and great submarine trenches result. Some of these ocean deeps are more depressed below sea level than the Himalayas are raised above.

Earthquakes rattle the plate margins, shallow where the plates rub near the margins, deep where the underlying plate is shoved down into the slushy depths. Eventually, earth's heat melts the sinking plate. The molten lighter elements become activated and press upward like the gas bubbles in a soft drink. They rise to the surface and punch through the

Titus Canyon, southern California. The forces resulting from plate collision may be transmitted great distances and crush rocks hundreds of miles from the plate borders. Some Great Basin rocks bear the scars of ancient plate conflict. *John Running*.

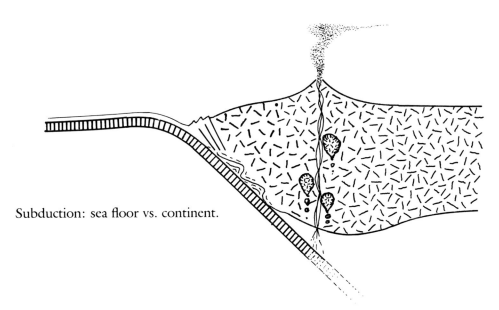

Subduction: sea floor vs. continent.

overlying oceanic plate several hundred miles seaward from the continental margin. Volcanic islands result. Hold a knife blade at forty-five degrees to an apple and press it in an inch or so. The result is a gracefully curved slice through the skin. The same arcuate curve is found where volcanos pierce the spherical crust over plunging oceanic plates. Long arcuate island arcs, crowned by volcanos, are the surface testimony to the melting of an ocean plate. Thus Japan, the Aleutian Islands, and Indonesia are such recycled ocean floors now rejuvenated into island arcs.

If a continent collides against a sea floor, the heavy ocean rocks almost always slide under the continent. The continental margin is subjected to compressive stresses and breaks into great sheets that slide atop each other in great stacks, as if they were lithologic pancakes. The huge pile is folded upward in convoluted wrinkles, and mountain chains form. The descending oceanic plate melts and portions recycle to the surface as volcanos. There is a significant difference, however, between these volcanos and their oceanic counterparts. In a continent–sea floor collision, the rising magma punches through the continent. It melts part of the lightweight continental rocks and absorbs them. Magmas contaminated by continental rock are silica-rich, creating a sticky and gooey lava. The plumbing of these continental volcanos becomes clogged with siliceous gunk. High pressures below press upward against the viscous plug. Spasmodically, the pressure blasts an explosive opening to the surface, taking a flank of Mount St. Helens or cities such as Pompeii or St. Pierre with it.

Continent against continent collisions originate in a subtle way. They usually begin under the disguise of a sea floor–continent collision. The sea floor plate subducts under the continent. A second continental block, fused to the sea floor, slowly advances as the oceanic rocks slide under the original continent. The ocean narrows as the massive bergs approach. Finally, the ocean closes and the continents crush together. Equal densities make equal opponents, and neither will yield to the other. Like two Japanese sumo wrestlers, the ponderous masses crush atop each other. Great sheets pile up, fold together, and fuse into a sutured scar of the former ocean. Huge earthquakes—like the one in 1976 that killed hundreds of thousands in Tianjin, China—attest to the titanic forces involved. The Himalayas top the earth due to such a collision between India and the soft underbelly of Asia. Mountains can grow no higher on earth than Everest. The weight of the piled-up rocks presses the mass into

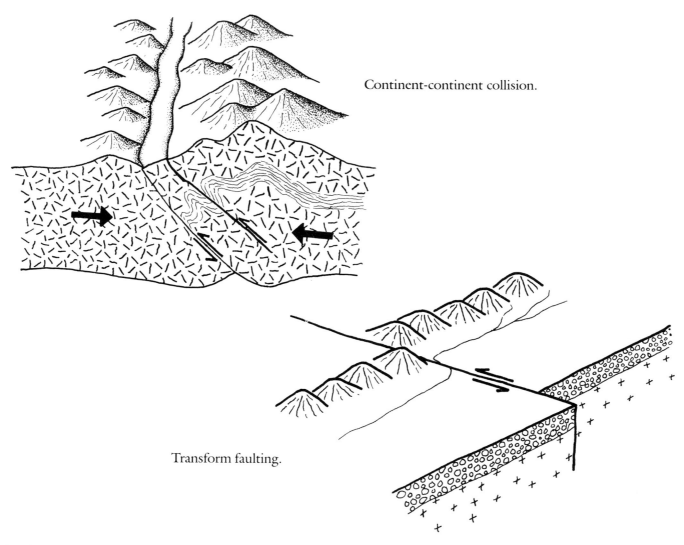

Continent-continent collision.

Transform faulting.

the depths, and it melts from the bottom as fast as it piles up above.

Sometimes plates slide past each other. This is no passive affair, as each plate elbows for more room. Great fault zones such as the San Andreas, Altyn Tagh, Denali, and Trans-Alpine faults mark such boundaries. Zones of pressure slowly build and snap like stretched rubber bands. Many such faults rip both sea floors and continents along side-swiping plates.

What happens to old sea floors? They slide under other plates along collision boundaries. They are digested and recycled at depth. When plates move apart, they make oceans. When they move together, they create mountains, studded by volcanos.

We have lungs, not gills, because of plate motions. The erosive force of water would have leveled the earth below sea level long ago. Tectonic plates, powered by the differential heat of the earth's interior, lift rock back above the ocean. Water erosion and plate motions—the interplay between these forces of destruction and uplift—create the topography of earth. We live on a giant see-saw, always seeking equilibrium. No other planet we know is like our unique home.

8 Shaping of the Land

THOUSANDS OF YEARS ago, Chinese sages developed a philosophy based on careful observation of nature. They concluded that everything is in a dynamic balance of opposites. Nothing is absolute or final. Modern scientists agree. The supposed absolutes of science, careful scrutiny and blunt honesty, yield to dynamic equilibria. For geologists, uplift and erosion are the yin and yang of earth processes. The forces work against each other in a conflict that has continued for eons and in which neither has overpowered the other.

The Taoists also observed that the most yielding of materials, water, can destroy the most rigid, rock. Indeed, today we recognize that of all the agents of erosion, none wields more destructive tools than water. Simple in chemistry and deceptively compliant, water comes in great variety. Disguised as ice or as clouds, underground or in streams and rivers, in sparkling crystals or in globe-circling oceans, water is always the simple and the yielding. Water is the destroyer and the creator of rock. Dynamic opposites become complements.

EQUILIBRIUM

To a granite, home is a hot, high-pressure environment. Bury a granite under the crushing weight of tens of thousands of feet of rock and granite is at home. Heat the subterranean pressure cooker to thousands of degrees and it is in equilibrium. This is the natural habitat of a granite.

Remove the granite from its home, and like a tourist in a foreign land it must make some radical adjustments. At the land surface, the feldspars and micas become clay. The quartz becomes sand grains. Granite no longer bears any resemblance to its original self. If I were to remain in Timbuktu long enough, you probably wouldn't recognize me either.

One of the fundamental simple facts of nature is change. We are well aware of the changes we observe in the biologic world around us. Because the life span of a rock so exceeds our own, we tend to overlook the changes in the rock. They exist. All rocks are in the process of change. They must all reach equilibrium with the environment in which they find themselves if it is different from their native environment. The earth is big. Few rocks are native to the thin skin at the surface. These changes result in weathering, and the movement of the by-products of weathering results in erosion.

WEATHERING

Rocks rot. Most of us don't think of such a thing. Rocks seem permanent, and most seem to be solid. They are, after all, our foundation. But ask a mountaineer, or anyone who scrambles on rocks for fun or for a living. One learns to place little faith in untested rocks. Chemical and physical processes alter the original mineral character of the rock, and it disintegrates. Nothing is permanent.

To attain equilibrium requires change. Few rocks were originally deposited or formed into rocks under surface conditions. Therefore, most rocks must change. The different chemical nature of the earth's surface and subsurface requires the rocks to experience chemical change. Since the surface is rich with oxygen, carbon dioxide, and water, most of the changes involve these chemicals. Most rocks are softened by these changes.

There are also profound differences in the physical nature of the land surface compared to the pressures and temperatures at depth. Rocks find relief at the surface and often expand. Layers often peel off. There are also sudden temperature changes. Ice expands in cracks and pores, and the rocks are wedged

apart. Plants attack the rocks for a roothold. Minute organic fingers pry deeply into any weakness and thrust rock apart as the rootlets expand. Plants are destroyers of rocks. Rocks, on the other hand, help sustain life. The disintegrating minerals release nutrients for the plants.

Rocks weather. Therefore nothing, not even solid rock, can withstand the changes wrought by differing environments through time. Not even rocks can remain forever.

GRAVITY, THE PRIME MOVER

Physicists have been studying the forces that shape and drive the universe since the days when Newton dodged apples. They eventually differentiated four primary forces that can affect us all. During the past few years, with massive number-crunching computers, these forces have finally been resolved into a single primal and universal force: gravity.

Certainly in the down-to-earth field of geology, gravity is *the* force. Erosive energies are powered by gravity. Rocks fall, rivers flow, glaciers creep, all due to the invisible tug of gravity. The major force that motivates rock is gravity.

Less obvious, but equally as vital, are the uplifting forces powered by gravity. Heat differentials cause heavy, cold rock to sink under gravitational pull. Light, hot rocks rise due to the density contrast with the cold. Yin and yang—up and down—result from gravitational attraction, since without the down there would be no up. Erosion and uplift, then, are the two faces of the same force.

EROSION

Rocks don't move. I stared with disbelief at the cliff and the long snow-covered talus slope above me. The snow had muted any noise, and I had been engulfed in a silence so profound that I could hear my heart beat. Then there was sound. The syncopated muffled thumping of a rock hopping and bouncing down the slope. With hollow thuds, it jarred against other rocks and dislodged some. Soon the slope was alive with ricocheting rocks. Each left a zigzag trail behind in the snow. Drunken boulders off on a spree. I felt like a secret observer at a private ritual. I wondered how often this primitive rite was conducted, safe from the prying eyes of such ephemeral creatures as humans.

I thought back to other times I had seen the rocks move. Usually they are dislodged by the movement of animals. Many times I've been able to spot the elusive desert bighorn by the sound of a pebble jarred from a cliff face as the sponge-footed animal pranced from rock to rock. It is a rare occurrence when I've seen rocks move of their own volition, and I have been outdoors with the rocks much of my life.

There was the time when I was walking deep in a narrow canyon in a far recess of the Grand Canyon. A pebble whirred past my ear as it plummeted down from hundreds of feet above. It hit the rock floor of the canyon near my feet and powdered like a rifle slug slamming concrete. Only a white circle and dust over the dark surface remained on the rock slab. I looked around for other circles in the rock. Knowing what to look for, I saw them. Most were gray, not white. Time had muted them to a universal gray. There weren't many marks, attesting to the rarity of lithologic bullets. I heaved a heavy sigh of relief, still shaken by the close encounter.

Another evening was shattered by rocks on the move. I still shudder remembering. As the sun slipped behind the Henry Mountains to the west, the cool quiet of twilight began to glide across the canyonlands. The best time of day. I sat on a cliff edge reworking my day's field notes. I had not seen another soul for over two weeks. My mental and physical state more closely resembled an Anasazi than a twentieth-century scientist. The days of human silence and the heat of a Utah sun had left a deep mark on my psyche, but I was refreshed. I had found my pot filled with water from the day's worth of slow dripping from my spring, and my stomach bulged with excess fluid. Dusk, at the end of a day of field work, is refreshing. The quiet beauty filtered through me, yielding a total release from any tension. My feet dangled loosely off the rimrock, and far below a canyon wren bubbled its trilling call off the rocks. Crickets were warming up for the evening concert. Sunlight bathed the snow-covered Sierra LaSal to the east. Peace, after a hot day of climbing over and talking with the rocks.

The rifle crack caught me unawares. The enormous intensity of the ear-splitting sound fractured the silence. In my reflex reaction, I almost flinched off my perch on the canyon edge. With disbelief, I saw the canyon wall opposite me fall vertically dozens of feet. It crushed onto the underlying gravel slope, and then tilted slowly toward me. It leaned into the canyon. In slow motion, it seemed both then and now in my memory, the wall canted outward and fell as a great megaton sheet toward the floor. It shattered with thunder and disintegrated into house-sized blocks as it struck the slope below. Dust rose from the scene and hid most of the events that followed. An occasional bouncing boulder raced down slope ahead of the cloud. The thunder distantly echoed and was gone. Then, abruptly, there was silence. The dust cloud slowly filled the canyon and billowed past me on air currents generated by the falling rock. The dusk darkened the view. I sat until it was black night and stars poked holes in the velvet over my head. Rocks move.

Usually the moving of rock is slow and silent. Ice wedges rock up in the winter and gently drops it a microinch at a time down the slope in the thaw. Rain dislodges pebbles and they shift. Erosion is usually not dramatic, but it is inexorable. Every change in environment—every temperature fluctuation, every rain—and the land is reduced. With thousands of millions of years at hand, erosion doesn't have to be dramatic in a human life span. Time is on the side of the rocks.

WATER, THE MOST ACTIVE AGENT

"Water, water, everywhere." Water is found everywhere over, on, or in the earth. Almost three-fourths of the earth's surface is ocean, and ice covers a fair portion of the remainder. The history of life is bound to water, and even now those of us plants and animals that live above the level of the sea carry water in our cells and tissues. Almost all the water on earth was derived from the earliest volcanic processes of the earth. It came from rock. Water, the gift from rock, is, however, the prime agent in rock destruction.

No erosive force is as important as water. None even challenges its number one ranking. More than thirty thousand cubic miles of water fall on the land area of the earth every year. Such an annual deluge, pulled downhill by ubiquitous gravitational forces, is an incredibly powerful agent of erosion. However, even with all the effort involved in the turbulent motion of all the raindrops and rivers of earth, the balance between uplift from earth heat and erosion by water energy is exact. The equation, uplift equals erosion, is a perfect balance of the dynamic pendulum of earth processes.

Scientists who study rivers have evolved another natural equation for understanding how rivers work. The carrying power of a river varies to the cube of its velocity. Simply stated, for every doubling of a river's speed, it can carry three times more sediment. Double it twice, and it hauls nine times the debris. Double it again, and it will move twenty-seven times more material. The erosion of the earth is not so much done by the day-to-day wearing away through the pitter-patter of raindrops or the swish of a lazy river. It is the torrential rains and churning floods that might occur once in the memory of a human that scour the earth. The Great Basin area is well known for its flash floods. Here, in the desert region, there are few plants to bind the soil and rock. There may only be ten inches of rain in a year, but it may come in two or three storms. The dry washes swell with brown, churning, viscous flows as the earth is shaped by the cutting fingers of fluid.

Other matter-of-fact observations will help our appreciation of rivers. Rivers, except those within the Great Basin, flow into the ocean. No river enters the sea at an elevation lower than sea level. Any valley that lies more than a few hundred feet below sea level was not carved primarily by a river. Death Valley, the low point of the Great Basin, has the Amargosa River flowing into the valley, but no rivers flow out. Death Valley was not carved by a river.

The Funeral Range, Death Valley. All rocks, regardless of their antiquity or variety, are exposed to destruction at the earth's surface. Metamorphic rocks in the highlands and sedimentary rocks in the foreground are all disintegrating and changing. *John Running*.

Ubehebe Crater, Death Valley. Some earth processes are violent within the span of human lives. This volcanic crater was blasted by the expansion and force of volcanic gases. *John Running.*

On the contrary, a river carries sediment into the valley.

Rivers usually join at a common level, allowing canoeists to travel between the junction of a tributary and the main stream without going over a waterfall. Tributaries usually form with the acute angle upstream, so that a river system generally shows the branching pattern of a tree. The master drainage is the trunk, and the tributaries are the branches. Whenever we find exceptions to these observations, and there are many in the Great Basin, there is an interesting geological story to decipher.

These observations combine into a single unifying concept regarding rivers: base level. Base level is the lowest elevation to which a river can carve its valley. The ultimate base level is the ocean. The trunk stream acts as a local base level for all its tributaries.

The higher the stream is above base level, the more active will be the downward cutting. Conversely, those rivers near the base level are sluggish and slow moving. The headwaters of a river, high above the ocean, carve more intensely than the lower reaches near the sea. Some geologists have made a rather quaint comparison: streams high above base level are youthful. They become mature in their lower reaches, and ultimately old age sets in as they near base level.

But even an old-timer can cut a jig with a little uplifting. A raised land such as the Great Basin will acquire a new relationship to base level. It can be vigorously eroded again. Deep canyons, rugged peaks, and great elevational relief are signs of such revitalization. Even a rejuvenated region cannot erase all signs of age, and the wrinkles of the land will reflect the superposition of the renewed vigor atop the old visage.

It is base level, then, and its relationship to all waterways that order and control the processes of stream erosion. Since streams are the prime movers in the work force of erosion, the appearance of our landscape is decipherable by understanding base level.

If rivers only cut down vertically, their valleys would be just as wide as the river. Such narrow slots do indeed form in very hard and uniform rock or when there has been recent uplift and rejuvenation. Most rivers, however, occupy broad canyons or valleys. The side-cutting process is descriptively called mass wasting. Gravity, again, is the prime mover in mass-wasting processes. Water freezes in cracks and wedges rock fragments free. Cliffs collapse thunderously, but rarely, along vertical fractures. Rains wash small particles down the slopes. If the flicker of geologic time were accelerated, all slopes would be in continuous movement. The land surface would be continuously creeping downhill. The debris is carried down the slope through mass wasting. Below lies the river, the conveyor belt that unceremoniously hauls away the continental garbage to the sea.

ICE, THE HIGH MOUNTAIN MODIFIER

Only by lucky accident is the earth spared from being completely ice covered. It is just the right size to have enough gravitational force to hold water and keep it from spinning off into the void and just the right distance from the sun to keep the water mostly in a liquid state.

Even today, in these relatively mild times, ice covers more than 10 percent of the earth's surface—six million square miles of ice. An area the size of South America is blanketed by glaciers. If the ice were evenly distributed over all the earth, we would be buried under a layer four hundred feet thick.

What a different appearance the earth would have if we were either slightly warmer or slightly colder. Geologists, looking back through the shimmering mirror of time, can read the rocks and see the consequences of either more heat or more cold. Many times during both the recent and the ancient past, the earth has had climates radically different from today.

Only a few thousands of years ago, a blink in the long flicker of time, the northern part of our continent was buried under as much as two miles of glacial ice. Humans stared in awe as crystalline white rivers moved ponderously down mountain slopes.

Some few lived near the margins of continent-covering ice sheets thousands of feet thick. These great Pleistocene continental ice sheets scoured the land. At their edges and in their wake, great piles of glacial debris were left scattered helter-skelter over the landscape. The land will never again be the same.

The polar regions, or at least the high latitudes, are usually considered home for properly behaved ice. The climate of the polar areas is sufficiently frigid to maintain the deep freeze. But glaciers affect high mountainous regions far to the south of the poles. We tend to forget that tropical regions and even the equator may have glacier-clad mountains.

During the colder climes of the Pleistocene, some of the mountain ranges in the Great Basin, far to the south of the great ice sheets, were covered with ever-deepening mounds of snow. The accumulation built such a pile that the deeply buried snow was pressured into ice and began to flow down the flanks of the mountains. These mountain glaciers ripped at the flanks of the highlands, and the ranges were scratched indelibly deep. Recent uplift is the reason for the beauty of our jagged mountains. But the rugged nature of some of our highest Great Basin mountains is the result of having been uplifted into the high elevations of Pleistocene frigidity. Here, in the realm of ice, mountain glaciers ripped at the up-lifted rock and shaped it into some of our most impressive peaks. The Rubies, Wasatch, and Sierra all felt glacial scour.

There was a major side effect of Pleistocene glaciation and cooler climates in the Great Basin. The greater precipitation and decreased evaporation created greater runoff of water from the mountains to the lowlands. The residual sediment that washed from the mountains litters the valleys. Lakes filled many of the valleys. Etched shore lines are obvious throughout the region. The result of the Ice Age was not only the ice-carved mountains but also erosion by liquid water from the ice melt and ice-induced rainfall. Here the fluid of the Ice Age affected the entire landscape, and its scour is still evident.

GROUNDWATER

Water deep under the ground (out of sight and hence out of mind) is an active agent in modifying the earth. Groundwater dissolves rock, but it can also form rock through precipitation. Chemical controls affect the fate of rock. Water underground moves through the small cracks and pores of the rock. The tortuous path is convolute and has many dead ends. Groundwater moves with such slowness that a glacier would become impatient at its progress. It might move an average of five feet a year, and that's in a good year.

Most of its work is so subtle and unenergetic that there is virtually no real physical effect of its activity at all, only chemical. Groundwater disassembles the rock in an oblique manner by taking apart the material internally. This attack is mounted by infiltration.

The most vulnerable rocks are those that form under the conditions found at or near the surface of the earth. Sedimentary rocks are often barely rock, since they are frequently lightly cemented or pressurized sands, muds, or gravels. The glue that binds the fragments is usually lime or silica precipitated from groundwater into pores between the pieces. What the water gives, the water can take away, and it does. The disassembled remnants of once-sedimentary rocks lie around us everywhere, testimony to the industrious nature of groundwater.

Some sedimentary rocks are especially vulnerable to groundwater. These are rocks formed of chemical precipitates from the antediluvian seas or lakes. They are the materials that were once dissolved in water and are therefore soluble if water should find them once again. But in the Great Basin the dissolved minerals are blocked in their journey to the ocean. Water leaves the Basin by evaporation, leaving the soluble constituents behind. Salt, gypsum, and anhydrite are salty rocks of the desert valleys of the Great Basin. Winds blow dust and sand over the saline surface, covering the whiteness. Rains occasionally flood the valleys and dissolve the salts. But soon the water evaporates and the newly precipitated minerals lie fresh and white and clean once

Double Hot Springs, northern Nevada. Groundwater emerges from the subterranean passages through which it has moved for thousands of years. Hot springs are common in the Great Basin as groundwater seeps to depths where it is heated by geothermal energy. The return to the surface is usually along deep fault systems which intersect the slowly seeping waters. *Rick Stetter*.

Sand dunes, Death Valley. Winds blow the rock fragments into piles in the lee of hills or mountains. Sometimes winds swirl down canyons into valleys and trap sand in their vortices. Many valleys in the Great Basin have dunefields. *John Running*.

again. Shimmering in the high sun, salt pans stretch to the horizon in some of our largest valleys. You don't find such temporal materials lying around in the wet swamps of the Everglades in Florida or littering the foggy alluvial coasts of Oregon. Salts are the denizens of our drought lands.

Some rocks, such as limestone or dolomite, were originally precipitated from seas or biologically removed from sea water by clever organisms to make shelly defense shelters. These rocks are all more or less soluble. When they are buried, groundwater saturates them and solution begins. If the groundwater is moving, it carries away the soluble materials and leaves holes underground. The Paleozoic limestones of the eastern Great Basin are especially vulnerable to solution. Many ranges have solution cavities, widened fractures, and caves.

Caves are the delight of most of us. The silence of the stygian blackness is punctuated only by the muffled thump of our heartbeats and the light plop of a water drop falling somewhere in the void. The incredible beauty of stalactites, stalagmites, helictites, and a thousand other cave formations taxes our imaginations in their frailty and boldness of design. They are the delicate handiwork of groundwater.

OCEANS

The scattered rubble left behind by the titanic struggle between uplift and erosion ultimately is swept into the sea. There may be temporary repose in a lake bottom or in the intermontane valleys of the Great Basin, but in the time scale of the earth, rest is only afforded in the silent depths of the sea. Even here, snugly secure beneath thousands of feet of water, in the cold, black, quiescent bottom muds, the repose will also be ultimately disturbed.

The continued shifting of the great plates of the earth may uplift ocean bottoms above the sea. Many of the mountain ranges of the Great Basin are composed of oceanic sedimentary rocks. Here, exposed to the probing erosive fingers of water or the crystalline scratching of ice, they again become mobilized. Or the passivity of the sea floor may be de-

stroyed by lateral motion. As oceanic plates are forced under continents or other sea floor plates, the sweepings of erosion may be crushed against the continental margin and forced upward into the scouring world of sunlight. This has happened several times in the Paleozoic past of central Nevada. Sometimes the sediments are not uplifted above the sea but are instead carried by the suicidal sea floor plate to great depths below the continent. Here the erosive garbage is either metamorphosed and fused to the continent or melted and completely recycled. What was uplifted is eroded. That which was eroded is uplifted. "The result, therefore, of this physical inquiry, is that we find no vestige of a beginning, no prospect of an end," summarized James Hutton.

WIND

The Greek god Aeolus controlled the winds. Aeolian action today continues to modify the landscape. But a much-overrated god he is. Newcomers to our desert region see fantastic landscapes unlike anything they are accustomed to in the more humid parts of earth. Pinnacles, abutments, hoodoos, porticoes, mesas, gulches, and spires dominate the scenery. Unfamiliar with desert processes, overactive imaginations conjure obscure agents to be responsible for the weird forms. A passing visitor to the desert is often awed by the presence of wind. Gusty, unbroken, often intemperate, the aeolian forces are thrust upon the stranger in ways rarely encountered in the wetter regions. Wind in the humid lands is almost always accompanied by the overshadowing force of rain. Here, in the arid lands, the wind comes alone. Forceful and cutting, often laden with dust or sand grains that abrade exposed flesh or pit auto windshields, it appears to dominate the drought-scarred landscape and must surely be the primary erosive force.

Desert dwellers, however, know the real culprit. Even here, in the parched desert lands of the Great Basin, water is the primary erosive agent. Wind is only a gentle modifier, a light touch. A little sanding here, a modest carving there, and an occasional pol-

Valley of Fire, southern Nevada. Erosion molds and sculpts rock into forms which surprise and please the viewer. Internal weakness or pattern becomes the guide for external force to follow while breaking up the rock. *John Running.*

ish is about the extent of aeolian action. The hard rock resists the efforts of so gentle a force.

Water creates the tools for Aeolus. Detritus formed by the solution and scouring action of water is ultimately reduced to the size of silt and sand. Particles of quartz, harder than glass, resist further erosion and remain for eons. They may be carried by the interior drainages into the basins where they are slowly buried by younger detritus. Ultimately, these sediments may be converted to new rocks. Or, on the margins of the Great Basin, they may be washed out to the ocean where they may also become buried and cemented into rocks.

In the meantime, these tiny resistant grains are just the right size to be moved around by the wind. Now Aeolus has working material with which to mold and shape the landscape. This is material to bury valley floors and to shape into high, curving, sensuous mounds. The wind creates dunes of almost infinite variety and form. Throughout the Great Basin, in Utah, Nevada, and California, there are fields of sand dunes. The zephyrs touch these dunes and mold the surface sand into asymmetrical ripples. The typical desert landscape is created by the heavy hand of water, but the gentle massaging of the wind has the final touch.

High Rock Canyon, northern Nevada. Water has followed fractures to slice through the plateau and carve the steep-walled canyon. Emigrants bound to Oregon followed the deep cut to leave the low deserts. *Tony Diebold*.

Death Valley. Uniformity of rock materials produces erosional forms which are similar. Symmetry of external form reflects internal congruity. Slopes of similar angles and equally spaced gullies testify to sameness. *John Running*.

PART II
Great Basin History

Death Valley turtleback. Ancient rocks, from the venerable Prepaleozoic time, are uplifted by faulting and arched by pressure. Younger rocks, pulled by gravity, slide off the old rocks. The resultant mountain of dense rock defies erosion and forms a turtleback. *John Running*.

9 Prepaleozoic

ASTRONOMERS, peering into their telescopes and recording their observations on sensitive plates, discovered a strange phenomenon many years ago. Every remote galaxy and every far-off object in the universe appeared to be retreating into space away from us. Another perplexing observation showed that the farther away an object was from earth, the faster was its velocity away from us. Astronomers were still sensitive about the sun-centered versus earth-centered universe controversies of Copernicus' time, so it seemed preposterous that earth was the center from which all matter was fleeing. It was as though we had a cosmic plague.

These observations resulted from the recording of a Doppler shift of radiation coming from all the distant stars of the universe. We experience the same Doppler shift every day in our perception of sound waves. As a car races by us on the highway, the *vroooom* is high pitched as the car approaches and lower pitched as the car passes us. If a galaxy approaches earth, the hydrogen radiation emitted by the galaxy shifts toward the blue end of the color spectrum. If galaxies are retreating from us, the hydrogen radiation shifts to the red end of the spectrum. No observable distant galaxy is blue-shifted. Radiation from distant galaxies is universally red-shifted. All the far-off objects of the universe are in retreat.

The farthest observable galaxies are about six billion light-years distant. These distant light sources are retreating from us at about one-half the speed of light. Where are they now? We have no idea. The light emitted by those galaxies is six billion years old. We know where they were then, but then is far before the age of the earth. Perhaps, for a geologist, the locations today are not as important as the fact that we are seeing light that was emitted long, long

ago. We can see backward into time. For instance, if we were located on a planet about ten trillion kilometers away from earth, we would be about 120 light-years distant. If we had an incredibly high-powered telescope, we would peer into the heavens tonight in the direction of earth and perhaps we would be able to watch the Battle of Gettysburg as it happens. This is not a replay, or photographic account, but the real thing! All visual history is traveling through space away from the source object at the speed of light. Thus, when we look at the stars we are seeing them as they appeared at some time in the past. We are seeing history.

The most distant galaxies and stars record the most ancient events. The objects that are the greatest distance from earth are quasars. The farthest away are perhaps sixteen billion light-years in distance. These strange objects seem to be small, very bright, starlike objects. Some astronomers believe that they are the very bright cores of otherwise ordinary galaxies. With such antiquity, they are perhaps the remnants of the earliest matter of the very young universe. Studying quasars is a real journey in time. They may be the clues to the youngest matter we know. Astronomers estimate that there were about fifty times as many quasars 4.6 billion years ago when the earth formed. Perhaps there were a thousand times as many quasars in the universe about thirteen billion years ago, when the universe itself was young. There seems to have been a much greater density of matter in the young universe than in the universe we observe today.

What does this tell us about the origin of the universe? Can the observations that all matter seems to be moving away from us and that all matter was formerly more dense lead toward a theory of origin? First, we know that earth is not really the center from which all matter is fleeing. Actually, all matter is universally receding and moving away from all other matter. Astronomers, in a humorous analogy, like to compare the universe to the baking of a loaf of raisin bread. The raisins are all crowded close together in the dough, but as the bread rises and the universe of dough expands, they move away from each other. If you were a piece of yeast hitching a ride on any one of the raisins, all the other raisins would appear to be moving away from you. So it is with earth. Through time, the matter in the enlarging universe expands away from all other matter. The density of the universe decreases as matter spreads out into the void.

To approximate the origin of the universe, we need to mentally reverse the process of the expanding bread. If matter is moving away from itself, doesn't this imply that once it was jammed together in a small, extremely dense mass? If all the trajectories of today's stars, galaxies, and quasars were plotted in reverse, wouldn't they coincide at a point? If the answers are yes, then what initiated the outward recession of matter? Could it all have begun with a bang? Perhaps. The generally accepted theory of most scientists today is indeed the big-bang theory.

If there was a big bang, how long has it been since this incredible event? What is the age of the universe? It seems reasonable to assume that it must be older than the oldest object yet located in space—the sixteen-billion-year-old quasars. Astronomers project backward in time from these ancient objects using an interesting mode of analysis. They observe that the older the quasar, the faster it is moving away from us. In fact, when they plot age and velocity it is apparent that for every additional one million light-years of quasar age, the objects are retreating at sixteen kilometers per second greater velocity. Since the fastest velocity possible is the speed of light, this becomes the universal speed limit. Matter eighteen billion light-years away would be traveling at the universal speed limit. We are two billion light-years short of observing these primordial quasars, so we have no direct record of the oldest matter. Therefore, we estimate that eighteen billion years is the time of the origin of the universe, the moment of the big bang.

Cosmologists, those scientists who study the origin of the universe, have debated whether there has been just one bang or many different bangs. Perhaps matter slows down in its recession and then gravitationally collapses back into a dense ball, only to bang again. Present data indicate that the bang was a one-time event. The matter of the universe is doomed to expand ever outward into the far reaches of space. We will always move farther away from our distant spatial neighbors, and matter will continue its sojourn into ever more lonely space.

THE ORIGIN OF THE SUN

Enormous clouds of interstellar dust and gas exploded into space following the big bang. Gravity organized some of these clouds and caused their condensation into stars. As we peer into the past, we can see stars in various stages of evolution: clouds, stars with nuclear engines, burned-out stellar ashes, and reorganization of cosmic dust back into clouds.

Our sun, about five billion years ago, gravitationally condensed into a slowly rotating dense cloud of matter. As the cloud contracted, the rotation increased, flattening the cloud into a disk. Spinning skaters can increase their rate of spin by pulling their arms close to their bodies. The collapsing cloud similarly increased its spin, or angular rotation. The most central part of this spinning disk contained the greatest density of matter, which contracted into our sun. It may have taken about fifty million years from the initial collapse of the cloud before the density of the new sun became sufficient to light the nuclear fuel of its interior. It takes time to fire up a nuclear generator.

Astronomers estimate that the sun has gathered enough mass to burn its fuel for about ten billion years. Since it is about five billion years old, it has just entered middle age.

THE EARTH'S BEGINNINGS

Many scientific hypotheses and theological concepts have been presented to explain the origin of our planet. Scientists today are not in agreement, and they probably never will be. It is the nature of scientific investigation to question and to operate with multiple hypotheses. The ideas in vogue today, supported by today's evidence, are discarded when new data evolve as quickly as last season's Dior design. I believe that the most generally accepted idea today is one we can call the "cold-earth model."

As the initial cloud of interstellar dust condensed into the original sun, the outer portions of the rotating disk condensed into cold dust particles and small planetesimals. Gravitational attraction pulled the lumps together. The accretion of the dust and planetesimals formed the remaining bodies of the solar system. Earth is the third major lump out from the sun.

The smashing impact of accreting hunks of matter generated heat in the forming earth. As the earth gained mass, gravity crushed the particles together and formed still more heat. The youthful earth melted because of these forces, and it is still mostly a molten planet.

Only ten thousand years or so after the initial melting, the first skin of solid matter probably began to form as thin, low-density silicate scums on the surface of the liquid earth. Heavy nickel-rich iron sank into the depths of the earth. This early differentiation of the earth created a silica-rich surface and a nickel-iron core. The initial patches of crust must have quickly foundered and been consumed by the boiling cauldron of the more dense interior. We find no remnants of the primordial crust on earth today.

Residual gases may have formed an early atmosphere of hydrogen, helium, neon, argon, krypton, and xenon. Fortunately for our form of life, the earth's mass and, hence, gravitational attraction were not sufficient to hold these light gases and they escaped back into space.

The rafts of silicate scum would, with increased cooling, eventually begin to form a continuous crust. The fragile skin would have been torn and fragmented continuously in the early days by the seething magmas below. Even today the crust is shifting and cracking over the still restless earth. With the first appearance of crust, geologic time on earth began. All our present long-range geologic time measurements require solids in order to set the clocks ticking. We have yet to find what is unequivocally the earliest crust. It is unlikely we will find it, since the earth is an organism that is continuously consuming itself.

BUILDING THE BASEMENT

The past shrouds itself in mystery. The farther back you reach in time, the more difficult it is to find data. Information is buried or forgotten. The record becomes increasingly sparse since there are no spectators left and the records are lost through one catastrophic event or another. The factual record of your father would likely fill file cabinets, that of your great-grandfather perhaps a few pages, but that of your predecessor twenty times removed would probably have little more than a name and a few dates, if that. So it is with the earth. The earliest events in the Great Basin are unknown, since there are no records left. We know little about the Prepaleozoic since virtually all the remnants not destroyed by erosion are buried deep beneath layers of younger rock. But geologists are trained to speculate about what cannot be seen.

The earliest evidence we have comes from some Prepaleozoic metamorphic rocks exposed in the Grouse Creek and Raft River ranges in northwestern Utah near the Nevada-Idaho borders. These oldest rocks—about 2.5 billion years old—are schists. Even this old sequence of rocks is little more than halfway back to the origin of earth. If the age of the earth were compared to that of the average person whose life span is seventy-five years, these rocks would be equivalent in age to that of a thirty-three-year-old person, almost middle-aged. What happened here in the Great Basin during those first thirty-three years? We don't know.

What is the extent of these rocks? We don't know this either. There are some tantalizing clues to suggest that the rocks below southern Idaho and northeastern Nevada are also at least Prepaleozoic age.

The Great Basin 2 billion years ago. Shorefront property.

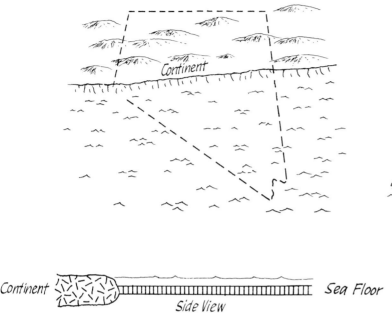

The Great Basin 1.75 billion years ago. Subduction begins.

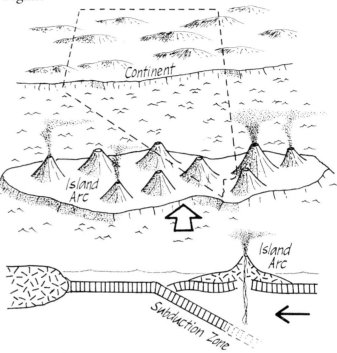

Magmas that intrude upward through oceanic lithosphere are fundamentally different from those that have poked up through continental rocks. These messenger intrusions have strontium isotope ratios that we think characterize the layers they penetrate. Mesozoic granites have pierced the crust below southern Idaho and northeastern Nevada. They reveal ratios that indicate continental crust underlying the surface rocks. The implication strongly suggests that they have penetrated Prepaleozoic rocks, perhaps as old as those in the Grouse Creek and Raft River ranges.

The ancient metamorphics that are exposed in these ranges bear evidence of plate tectonic movements. There are many granitic intrusions, widespread calc-alkalic volcanics, and metamorphic belts left behind as clues for geologists to use to reconstruct the shifted pieces of earth's puzzle. We interpret, with more guess than fact, that the moving plates had assembled together to form one massive continent out of the early crustal building blocks. Imbedded somewhere in the western portion of this supercontinent was the core of what would become North America.

In order to free the enclosed North American nucleus, the supercontinent had to fragment. We speculate that about two billion years ago rifting occurred. The split ruptured the land a short distance to the south of where the Grouse Creek and Raft River ranges are today. The fractured segment of the supercontinent slowly drifted off to the south. Young ocean basin rocks filled in the gap between the retreating margins of the segmented continent. The Grouse Creek and Raft River rocks were now close to the southern margin of the continental core of the North American continent. They were shorefront property.

Through the slow flicker of geologic time, new pressures built up in the convection cells that stir the earth's molten interior. The old directions of plate movement were redirected again about 1.75 billion years ago. The sea-floor plate to the south of the continent reversed direction and began to move northerly against the resisting craton. A new sub-

The Great Basin 1.748 billion years ago. Island arc accretion.

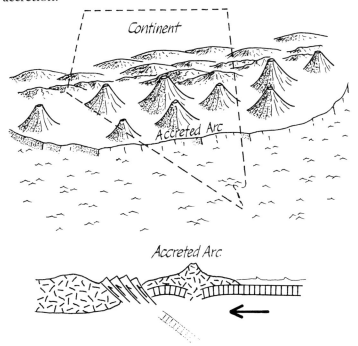

duction zone formed. The ocean floor remained firmly attached to the continent, but offshore it buckled and cracked. The northerly shoving oceanic plate overrode the sea floor fixed to the continent and a subduction zone formed. Ocean floor moved beneath ocean floor just to the south of the continent. As the accreted ocean plate slid under the northerly moving sea floor, it eventually reached depths where it melted. Molten lightweight components of the subducting plate intruded upward into the northerly moving plate. This recycled sea-floor magma broke through the surface of the crust. Slowly an offshore arc of islands, studded by smoldering volcanos, materialized. These islands expanded in size as great quantities of magma surged upward and hardened into rock. Quickly, the uplifted young rocks were attacked by erosion. They were reduced to a pile of fragments that lay like lithic aprons on the flanks of the island arc.

The basin that lay between the continent and the newly formed island arc was flooded by this fragmentary rubble. Some of the debris was scoured from the continent as well as the volcanic arc. Yielding to the massive pressures of plate movement, the volcanic island arc slowly advanced against the continent. The movement crushed the near-shore sediments that had been dumped into the closing basin. Eventually, the arc and its bulldozed oceanic sediments smashed against the continental margin. The collision sheared and thrust the sediments northward. The continental margin cracked upward from the pressures. Highlands were again attacked by erosion and floods of sediments washed off the uplift. The force of collision bonded the volcanic arc tightly to the old continent. With the addition of this island arc, the edge of the continent now lay to the south of what would be the Great Basin. What had formerly been ocean floor was now younger Prepaleozoic crystalline rock sutured tightly to the continent. The southern Great Basin had a foundation.

These newly added rocks are part of a broad east-west belt of younger Prepaleozoic rocks that accreted to the continent in the island arc–continent

Raft River Range, northwestern Utah. The oldest rocks
in the Great Basin are these venerable metamorphic rocks.
The light-colored intrusions of pink granite have been
dated at about 2.5 billion years old. *Bill Fiero*.

Raft River Range, northwestern Utah. The subdued topography reflects the eroded Prepaleozoic ancient rocks of the oldest piece of North American continent found in the Great Basin. *Bill Fiero*.

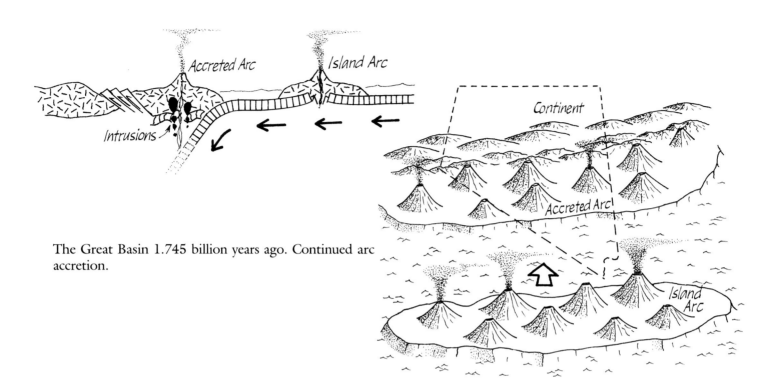

The Great Basin 1.745 billion years ago. Continued arc accretion.

collision. We aren't certain where the eastern limit of these rock provinces may be, but their far cousins may be residents of Wisconsin or perhaps eastern Canada. The westerly extent of the accreted belt is also unknown, since later north-south rifting sheared off the western region and carried away all the evidence.

Mother earth has few secrets that can't be ferreted out by scientific analysis. She can't even hide her age from the probing detectors of geologists. Radioactive clocks have been used to determine the age of these arc-accreted younger Prepaleozoic metamorphic rocks in the southern Great Basin. The age is revealed, through the counting of uranium-lead isotopes, to be 1.74 billion years old, plus or minus .25 billion years. The sampled rocks are the oldest rocks yet found in the southern Great Basin. Venerable indeed, but only a little older than one-half the age of the basement of the northern Great Basin. In human-earth terms these rocks are the equivalent of a forty-six-year-old person.

At about the same time as the arc-accretion, in-

trusive rocks forced their way upward into the crust of the southern Great Basin. These intrusives are unique, since they possess extra-large feldspar crystals. The distinctive rocks are termed rapakivi granite. Such rocks may be seen along the abutments of Davis Dam on the Colorado River and in the hills around Nelson, Nevada. Isotopic determination of the rapakivi reveals an age of 1.45 billion years old, plus or minus .25 billion years, or equal to a fifty-one-year-old human.

Thermal nudges from below the crust continued to jostle the plates after the accretion of the rocks of the southern Great Basin. A new subduction zone developed well to the south of the recently acquired continental margin. This time the northerly moving sea-floor plate slid directly under the continent and another subduction zone formed. Another volcanic arc formed offshore. Perhaps a small basin rifted open between the arc and the higher continent. History repeats itself.

Once again, the new arc advanced upon the continent, crushing the basin closed, and more material

Accreted Arcs

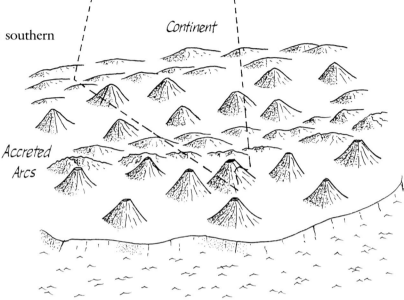

Continent

Accreted
Arcs

The Great Basin 1.74 billion years ago. The southern Great Basin is now part of the continent.

accreted to the continent. This suture took place about one billion years ago and lay well to the south of the Great Basin. The continent continued to grow in a southerly direction. The edge of the continent now lay to the south of the Texas border. This is the way continents are built.

The southern border of the North American continent did not reach completion until relatively recent time. In human-earth time spans, this was only sixteen and a half years ago, when the earth was fifty-eight years old. These basin closures and arc collisions resulted in the estimated accretion of eight hundred miles of new crust to the south of the Grouse Creek and Raft River rocks. This rate of crustal growth has been estimated to be about one-tenth of an inch per year, considerably less than the growth rate of your hair.

These ideas are hypothetical. Little evidence remains of these ancient events. But these ideas are as good as any that have been proposed. Much field work and considerable head scratching remain to be done. Perhaps these ideas will be totally revised to-

morrow. This is the way science works.

Regardless of the details of the hypotheses, the foundation of the eastern Great Basin, south of the Grouse Creek and Raft River ranges, is now in place. Much must be done to build atop this foundation before the edifice would be complete, but you can't build the house without the basement.

THE MISSING LINK

The youngest Nevada igneous basement rock, sutured on during arc collisions, is about 1.45 billion years old. The oldest Prepaleozoic rocks deposited above this basement in Nevada are approximately 650 million years old. The intervening 800 million years of Nevada history are a mystery. Virtually all the rocks are gone.

We have yet to find enough clues in the Great Basin to piece together the history of this missing piece of the Nevada Prepaleozoic, even though geologists have been diligently searching for many years. The gap exceeds all the time that has passed since the beginning of the Paleozoic, when the first shelled animals evolved, until the present. Surely this is sufficient time for monumental events to have occurred.

We don't even know why the evidence for this time interval is gone. Perhaps the ancient crystalline continent, born of collisions and volcanic outpourings, was forced high above sea level by the force of impact or the thermal bulging of the hot crust. Erosion would have been active in the ancient plant-free continental environment. Floods of debris, our missing Nevada rocks, would have been washed out to the oceans.

Elsewhere in the Great Basin there are scant shreds of evidence. Approximately 1.2 billion years ago, the constantly shifting cauldron of subcrustal heat beneath the Great Basin apparently generated a convecting current of especially hot magma. Great heat results in decreased density and the hot magma pressed upward against the resistance of the overlying crust. The crust bulged, fractured, and broke. Downdropped basins formed in the stretched, rigid crust, and sediments flooded into the troughs. Evidence of such early stretching and basin filling has been ferreted out of a few localities in the mountains of northern Utah and the Death Valley region of California.

There are modern analogies to these Prepaleozoic tensional fault-dropped valleys. Have you ever noticed the unique fracture patterns that crack a sidewalk when it is forced upward from below? The fractures often form three cracks that meet at a common point. Each angle is approximately 120 degrees. Brittle slabs, when forced upward from a single point below, often break in a triple junction. The brittle crust of the earth often behaves in the same way. When upwellings of heat force the brittle crust upward, it usually breaks in the classic three-way split. The Red Sea–East African rift valleys–Gulf of Aden region is a modern triple junction.

When continents are rifted, they often fail along such triple junctions. Usually two of the arms become the most active and join to become the fracture along which the continent ruptures. The third fracture then becomes inactive. These failed arms of triple junctions are termed aulacogens. In the East African region, where the Gulf of Aden and Red Sea arms are actively separating Africa from Arabia, the aulacogen is the great rift valley of East Africa. The lower Mississippi River valley is thought by some geologists to be a failed arm resulting from the breakup of North and South America. New England's Connecticut River Valley may also be an aulacogen that formed from a triple junction when Europe divorced North America.

Somewhere between 850 million and 1.2 billion years ago, major continental rifting began again. The continent, so painstakingly pieced together in earlier times, ruptured. The Great Basin tore along an arcuate set of fractures running roughly north-south through central Nevada. Aulacogens probably extended into the continent along the failed arms of triple junctions. The central portion of the main rift downdropped so far that the ocean flooded into the rift zone. The inexorable pressures of underlying thermal bulging slid the continental blocks off the bulge, widening the ocean basin. The great north-south rift sheared off all the continent that lay to the west. The adjacent uplifted continental margins were exposed above sea level as a mountainous highland. Slowly the western fragment or fragments drifted away from North America. A new ocean floor formed in the rifted gap. The western edge of North America ran north-south right through what is now the Great Basin, and we had ocean frontage.

How much of the continent drifted away? Where are the pieces today? We don't know. Speculation, founded on some evidence, places the largest piece in Siberia, but a definitive reconstruction has yet to be made.

In today's Great Basin, the western edge of the tear is buried by more recent sediments and volcanics. Its location is revealed, however, by those messenger granites that carry the tattletale strontium isotope ratios. A line following a north-south arc through central Nevada separates those intrusions that have penetrated continental lithosphere from those that have moved upward through oceanic rocks. Everything to the west would presumably

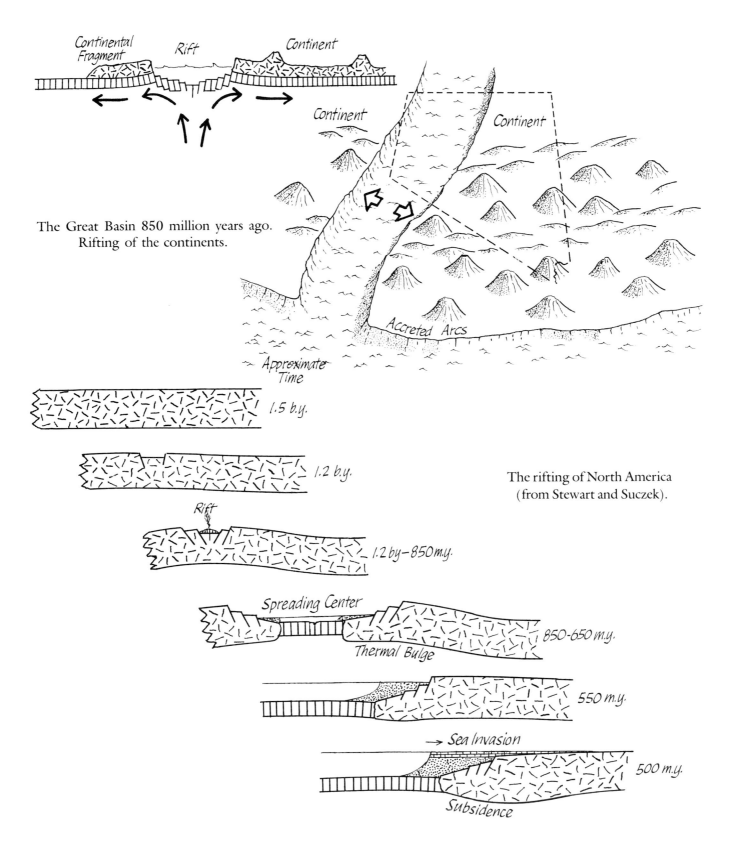

Continental Fragment

Rift

Continent

Continent

Continent

Accreted Arcs

The Great Basin 850 million years ago.
Rifting of the continents.

The rifting of North America
(from Stewart and Suczek).

Approximate Time

1.5 b.y.

1.2 b.y.

Rift

1.2 b.y.–850 m.y.

Spreading Center

Thermal Bulge

850–650 m.y.

550 m.y.

→ Sea Invasion

500 m.y.

Subsidence

Devil's Gate, central Nevada. A mountain range resulted from the accretion of plates in the central Great Basin 350 million years ago. Today, only the fragments of erosion from this range remain. Beds of this debris outcrop at Devil's Gate near Eureka, Nevada. *Tom Brownold.*

be underlain by younger oceanic lithosphere. The younger overlying rocks also clearly show the mark of these oceanic environments by their depositional patterns and fossils.

The stage is now set. The crystalline foundation is completed, and rifting has carved a new western continental borderland. The Great Basin is to be center stage for some exciting screenplay. If you are looking for geologic action, it is most exciting along the edges of continents.

10 The Paleozoic

PASSIVE SUBSIDENCE

Slowly, ponderously, the former piece of North America moved westerly. This huge cratonic fragment may have drifted across almost one-half of the globe before affixing itself onto the continental margin of Asia. Later fragments were glued onto its outboard margin, embedding the misplaced fragment of America deeply into Siberian exile. Meanwhile, back along the newly rifted border of North America, the thermal welt also moved westerly away from our shore. Our margin cooled, contracted, and subsided.

Just as a cooling oven loses the most heat in the first few minutes after it is turned off and slowly loses the remainder, so it is when magmatic heat is turned off. The initial heat loss, after the thermal spreading center moved away from the continent, was great, and consequently the contraction and sinking were greatest in the first years of cooling. Slowly, equilibrium was reestablished and the rate of thermal contraction slowed. Data from a more recent rifting episode in the Atlantic Ocean indicate that contraction episodes require about fifty million years.

The uplifted Prepaleozoic continent was barren. There were no land plants to hold the weathered soils, nor were there any land animals to witness the erosion of the highlands. The life in the sea was simple and so soft as to leave almost no record. Some glacial deposits from six hundred to seven hundred million years ago have been described in the western United States. Perhaps the earth was considerably cooler due to an elliptical orbit around the sun. Or maybe the highlands were elevated far enough to catch snow. It has also been proposed that rapid drifting may have brought the glaciated areas into the high latitudes during this one hundred-million-year time period, although this seems doubtful.

A great flood of sediments cascaded down the desolate flanks of the thermally swollen mountainous borderland and poured into the marginal sea. The dual forces of thermal contraction and erosion slowly lowered the continental highland. As the mountains were reduced, the erosive products decreased and the sediments piling up in the ocean along the continental border decreased in volume. The mountains were reduced to rubble and the continental margin was buried by a huge pile of debris. What had been a lofty highland was now a great westward-thickening prism of Prepaleozoic sediments. This was the beginning of the Cordilleran miogeocline, a slowly subsiding and sediment-filled trough adjacent to the continent. The clastics blanketed the near shore and continental shelf environments. Fragments of the eroded continent were swept by ocean currents as far west as central Nevada.

As thermal contraction continued, the continent sank beneath the sea. The ocean invaded easterly, across the subsiding crust, and shallow waters encroached over the continental domain. By Early Cambrian time as much as twenty thousand feet of clastic and carbonate sediments were deposited along this continental shelf along the edge of the craton.

The depositional environments from the late Prepaleozoic through the early Paleozoic were essentially the same. The rocks, therefore, are very similar. Geologists will stand on an outcrop and argue for hours, trying to determine the contact between the Prepaleozoic and the Paleozoic, unless there is an abundance of unequivocal fossils.

Jackson Mountains, northern Nevada. The black exotic rocks uplifted in this range were accreted to the North American continent during latest Paleozoic time when Sonomia collided with our continent. *Bill Fiero.*

Lincoln County. The central and eastern Great Basin were covered by a shallow sea during most of the early Paleozoic. These carbonate rocks, uplifted and tilted by more recent geologic events, were deposited in that ancient sea. *Bill Fiero*.

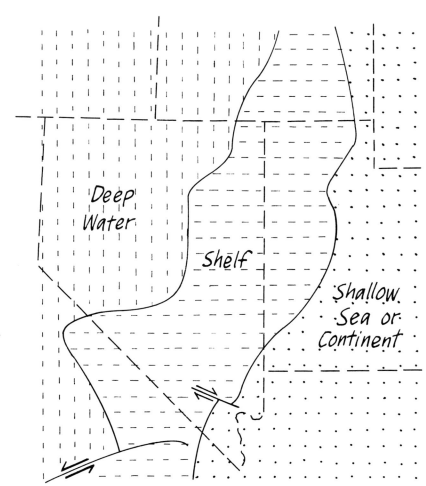

The Great Basin during early Paleozoic.

The sea was shallow in the eastern Great Basin. Along the shoreline we see today, mudcracks indicate periods of drying above sea level. Some of the muds have the distinct imprints of raindrops. Imagine holding a rock in which the impression of an antediluvian deluge are still preserved after six hundred million years! Even blasé geologists, often inured by constant exposure to the vast geologic time span encapsulated in the rocks around them, can't help but be impressed by the delicate preservation of such a fleeting moment of time preserved in a rock at least ten million times their age. The delicate tracings of the impressions of algae, which require sunlight for survival and hence live in shallow, well-lit waters, are also found in these rocks. Move-

ment of sediments by ancient tidal currents is indicated by the layering and ripple marks found in rock. The events of the past are similar to those we see today. Nothing is new.

In the central Great Basin, farther from the sediments washing off the continent to the east, the shallow waters glistened clear and sparkling in the warm tropical sunlight. The equator at this time probably trended approximately from Houston, Texas, through Hudson Bay, Canada, and our Great Basin was in the southern hemisphere.

Few clastic sediments washed out far from the shore. Lime precipitated out of the clear warm waters, ultimately hardening into limestone and dolomite. The shallow warm-water seas of the Cambrian were an excellent environment for pelecypods, trilobites, echinoderms, and algae. Fossil remains of these creatures are abundant, especially in the central and eastern Great Basin. These carbonates mark the transition between the clastic deposits to the east and the deeper oceanic environment of the west.

To the west of the shelf lay the deep ocean. Rocks of abyssal origin are rarely thrust onto the land. They are usually digested in subduction zones beneath continents. There are, however, outcrops in the western part of the Great Basin that may contain Cambrian abyssal rocks. These outcrops are suspect, however, since they have been moved eastward along later faults. Their original site may only be conjectured.

During the remainder of the Cambrian time, the shoreline continued to advance eastward until the beach zone lay from four hundred to more than six hundred miles east of the Great Basin. Much of the western United States was a shallow sea. The highlands were gone, and the sea flooded over a continent of low relief in a setting similar to that of the late Prepaleozoic and Early Cambrian. The rate of sedimentation in the miogeocline was incredibly slow as the shelf passively subsided. Freed from the rain of sandy clastics, the shelf was an area dominated by tide flats, lagoons, carbonate banks, reefs, and beaches. The rocks of this eastern or carbonate assemblage reflect these environments. The limestones of the shallow lagoons were often impregnated with magnesium solutions and turned to dolomite, the kissing cousin of limestone. This paleogeographic pattern was to continue until the Devonian.

During the Ordovician, the equator is thought to have extended from Big Bend, Texas, through North Dakota. Continental movement slowly drifted North America to the northwest. By the Late Devonian, the equator was on a line approximately from Los Angeles through the Great Basin to western Minnesota. Throughout most of the Paleozoic, therefore, our region was one of warm tropical climate. Our oceanic rocks basked in equatorial waters.

SNOWLINES AND TURBIDITY FLOWS

Scientists diving deep into the sea in submersibles have reported a strange phenomenon that seems to be more akin to the tales of mountain climbers than to marine scientists. They report that around each island or every continental shelf there is a "snowline." In the shallow water around the island or shelf, a fine white powder covers the sea bottom and its rocks. This powder, or "snow," is composed of minute precipitated granules of limestone. As the submersible dives deeper, the snow disappears along a rather distinct boundary, or snowline. Just like the mountains, the oceanic highlands are covered with snow. This observation relates to the strange solubility of calcium carbonate, or limestone. Unlike almost any other substance, limestone is more soluble in cold water than in warm water. What a paradox. Usually you heat water to get minerals to dissolve, but not so for limestone. Thus, a teapot accumulates limey deposits in the Great Basin where our drinking waters are saturated with limestone and the warming of the water causes carbonate to precipitate. Cold deep waters of the ocean are abrasive to limestone and it dissolves.

Silica behaves in a more customary fashion. It is less soluble in the cold deep water than in warm water. Hence, the oceanic deeps are the domain of siliceous materials. Deep-living marine plankton, the drifters of the sea, build their shells out of silica to resist the solution of cold water. When they die, there is a constant rain of their shells to the deep. Much of the sediment that accumulates in the black, frigid ocean bottom is composed of the last earthly remains of these plankton with silica shells.

Nonorganic siliceous sediments also litter the deep ocean floor. How does this material reach the abyss? Some is undoubtedly blown out to the open ocean by strong winds and settles down into the sea. Volcanic ash is dispersed worldwide by the winds. Silica-rich sediments also pour off the continents in great turbid flows, especially off the mouths of major rivers. I have vivid memories of flying across the Yellow Sea and observing swirls of suspended muds being carried hundreds of miles out to sea off the mouth of the Hwang Ho. Such erosional debris piles up in thick, unstable mounds on the subsea flanks of the continent. Triggered by earthquakes or by the sheer weight of the mounds of debris, the sediments may suddenly start to slide down the slope. The suspended sediment creates a turbid mass of water that surges down the slope driven by its own weight. Similar to an avalanche or landslide on the continents, this subsea turbid flow churns down the slope at great speed. Turbidity flows have engulfed divers and submersibles. They have snapped submarine cables. They also sweep an incredible volume of fine siliceous sediments into the deepest parts of the sea.

DEEP WATER: THE SLOPE

Back in early Paleozoic days, the siliceous erosional remnants of the highlands moved as turbidity flows down the continental flanks to the west of the carbonate-rich shelf. As a result, in the central Great Basin, in the region that was formerly the continental slope, there is a mixture of siliceous shales and carbonates. These rocks are a transitional sequence between the carbonate shelf and the siliceous deep ocean.

The cold waters of the early Paleozoic oceanic depths, westerly from the continent, dissolved out the carbonates. Only siliceous sediments and volcanics were stable at great depths. This sequence, exposed in a few ranges of the central and western Great Basin, is referred to by geologists as the siliceous or western assemblage.

Turbidity flows and submarine volcanic ash dominate, although a few carbonates attest to short periods of warmer or shallower water conditions. The volcanic ash was perhaps airborne from the west and fell into the sea. Or perhaps the ash was rafted in on mobile oceanic crust. The oldest unequivocal deep ocean deposit yet found is Ordovician. We are not certain about the thickness of these western rocks since later earth movements have greatly obscured the record.

This pattern of eastern carbonate, transitional, and western siliceous assemblages is plastered over the older, dominantly terrigenous late Prepaleozoic and Early Cambrian rocks. This pattern continued almost without interruption until the Devonian.

There were no major wrinkles in the Great Basin during the time span from the erosion of the Prepaleozoic highlands until Late Devonian. The sea floor was an almost featureless plain, angling down through the blackness of the sea toward the abyss. There are a few areas in Nevada that have been described as having some folding and warping, particularly in Late Cambrian and Ordovician time, but there is geological debate about the nature and extent of such tectonism. If it existed, it was probably of local importance. The first evidence of widespread tectonic activity is Late Devonian.

There may, however, have been some movement far off to the west of the continent during the time of continental quiescence. It is possible that the sea floor plate built up pressure from earth's internal heaving and pressed against the continent. Perhaps the thin oceanic plate broke somewhere far offshore. It may have begun to slide under the remnant sea floor that was still glued firmly to the continent. Such subduction carried the moving plate well below the remnant, where it melted. The light-weight molten components were out of balance surrounded by high-density deep magmas. They punched back to the surface through the overlying oceanic plate. If these events occurred, this may have created an offshore arc of volcanic islands. Between the hypothetical island arc and the continent there was an oceanic basin. If this arc existed in the early Paleozoic, it must have been far to the west of the Great Basin since western assemblage oceanic rocks have no pre-Silurian arc-derived volcanic components. Siluro-Devonian deep-water rocks in Nevada may contain significant volcanic rocks, but this evidence is also debated.

Another scenario is possible. Perhaps if there were a western landmass, it would have been a continental fragment that drifted across the ocean toward North America. Such a plate would not neces-

Schell Creek Range, Nevada. Thin-bedded siltstones from the deep ocean are pushed up to the surface by immense faults. The faulting resulted from plates colliding in the central Great Basin. *Bill Fiero*.

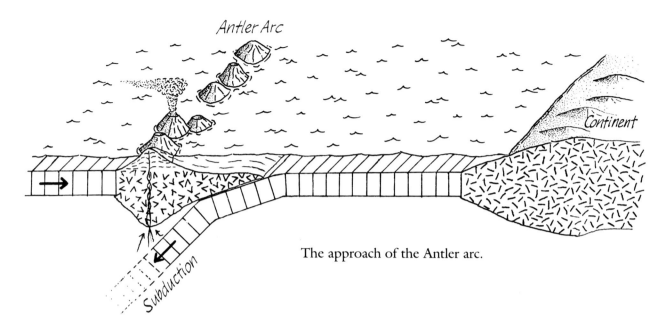

The approach of the Antler arc.

sarily have had active volcanism. Some of the early Paleozoic deep-water western assemblage clastics may have been eroded off this continental landmass rather than an island arc. The siliceous sediments would have swept into the oceanic basin lying between the approaching fragment and the continent. By Devonian time, however, most geologists believe there is sufficient evidence to establish a landmass lying offshore from the continent. Most probably, this was an island arc.

THE ARC COLLIDES

For three hundred million years, the continental margin was passive and the general pattern of shelf and slope sedimentation remained essentially unchanged. Then, suddenly, in Late Devonian the pattern of the Cordilleran miogeocline came to an abrupt and dramatic close. The passive continental margin suddenly became involved in large-scale compression and mountain building. Where there had been a shallow sea with a gently sloping undeformed sea floor, the Antler Mountains would rise.

Large-scale episodes of tectonic movement that create great uplifts, warp up mountain ranges, or generally rearrange pieces of the crust are called orogenies by geologists. The Antler orogeny differs from other orogenies. The forces that cause mountain building usually generate great pressures and heat. Orogenies are usually accompanied by metamorphism of rock and by igneous intrusions into the contorted rocks. The Antler has no known metamorphism or intrusion.

The siliceous and volcanic rocks of the western assemblage, which were derived from the erosion of the continent and perhaps partly shed from an offshore landmass, had blanketed the deep-water basin that lay to the west of the continent. Squeezed easterly by great compressive forces, these western rocks yielded in sheetlike faults, forcing young deep-water rocks over old slope or shelf sediments. The young rocks sometimes squeezed between the layers of the old rocks. The stack of previously almost horizontal and carefully layered oceanic rocks were shuffled like a deck of cards. The sorting out of the suits has hardly been accomplished by geologists diligently working on the problem for over fifty years, let alone discrimination between the aces and eights.

The fault slices, riding over the top of one another, were thrust eastward at a rate of about one-

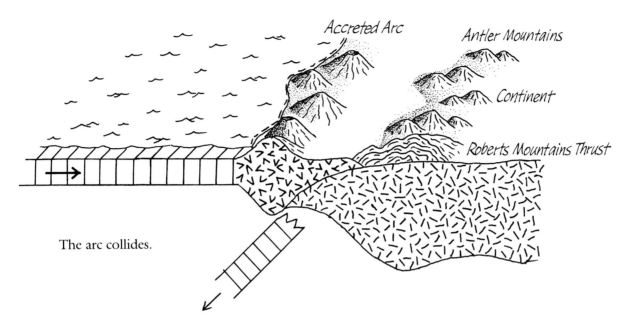

Accreted Arc

Antler Mountains

Continent

Roberts Mountains Thrust

The arc collides.

half inch per year. It took about eight million years for the western assemblage to override the eastern early Paleozoic carbonates by about ninety miles. This great fault has been called the Roberts Mountains thrust.

What caused this monumental event? We don't know. A recent conference of fifty-one geologists discussed ten possible models for the Antler orogeny. Perhaps the most straightforward explanation is that the volcanic arc collided with the continent just as another had far back in Prepaleozoic time. This time the arc arrived from a westerly direction. Perhaps the arc had been patiently lying offshore throughout the early part of the Paleozoic, or maybe it had been built by subducting processes in the Late Devonian. As it moved ponderously against the continent, the deep-water sediments of the oceanic basin would have been crushed between the easterly moving arc and the continent. As the jaws of the vise closed, the sea floor sediments would thrust eastward in great fault slices.

The result of the collision is evident throughout many of the mountain ranges of the central Great Basin. Early Paleozoic rocks are deformed, and deep-water rocks lie above or below their synchronous

shallow-water counterparts. We may debate the cause, but there is no geologic doubt about the existence of the Antler orogeny.

There is one major problem with the island arc–continent collision hypothesis to explain the Antler orogeny. The arc has never been found. There are no rocks yet described that are incontestably from the arc. How do geologists explain a mountain-building event caused by island arc–continent collision when there is no evidence of the prime cause, the arc? Simple. They merely eliminate the arc.

An interesting but unproven speculation might account for the missing arc. This hypothesis would suggest that the continent ripped open along a jagged two-armed rift that ran roughly north-south just to the west of the former Antler belt. Deep aulocogens, representing the failed arms of triple junctions, may have sliced deeply into the torn continental border. The earlier suture zone became unstitched and the cryptic arc drifted back out to sea. A new oceanic environment, replete with volcanic sea floor, developed off the western margin of the continent. Some geologists hypothesize that this ocean may have become as wide as the present Atlantic Ocean. This is a convenient way to dispose of a secretive island arc. At the close of the Paleozoic, this arc may return again to cause another mountain-building episode. But that is another story.

An alternative scenario could be proposed. Another way to get rid of the arc is by thermal subsidence. As the accreted arc cooled, it would sink. The sea would encroach over the arc rocks and they would be buried by erosional rubble from the continent. Or, following a variation on the theme, perhaps rocks of a later collisional event would ride over the subsided Antler arc rocks and hide or destroy them in a subduction zone. The evidence of the Antler arc would be neatly shoved beneath the carpet.

Which hypothesis is correct? Frankly, we have too little evidence to say. All we know is that there was a major rock-busting event. We think it may have been caused by arc collision but the arc hasn't been found. Maybe it rifted away or maybe it subsided and was buried. Perhaps it subducted beneath the continent. Perhaps all these ideas will be revised next week by a lone geologist chipping away on a rock in the Great Basin. The debate about the origin of the Antler orogeny is current. The final story is not yet written on this remarkable tectonic event and perhaps it never will be.

How do you date a 350-million-year-old orogeny? With a little diligent field work. The easiest way is to determine the ages of the youngest strata moved. This gives you a maximum date. If you then date the oldest rocks to blanket the uplift, you have a minimum date. Thus, with such a span, you have bracketed the date of the orogeny. The closer the maximum and minimum dates are to each other, the more accurate the dating of the mountain building.

The youngest rocks involved in the moving block, or allochthon, of the Roberts Mountains thrusts are the latest Devonian, and the oldest rocks to cover the thrust sheets after the orogeny are the latest Early Mississippian. Thus, the orogeny occurred within a relatively short interval in Late Devonian and Early Mississippian time. The allochthon probably moved several centimeters a year to cover the ninety miles in such a short time span.

There are holes, or erosional windows, in the allochthon. Termed fensters by geologists, they reveal the underlying eastern or transitional assemblages and substantiate the existence and extent of the fault plate. The Roberts Mountains thrust is clearly evident in northern and central Nevada, but the location of the allochthon is less certain to the south, where the exact relationships of the western and eastern facies are not clear. Some geologists have extended the allochthonous block, on the basis of scattered rock outcrops, as far as western California.

THE POST-ANTLER WORLD

Nestled in a small stream valley carved into the mountains of central Nevada is the old mining town of Eureka. A once-thriving community that boasted an opera house, a railroad, and a lively newspaper, Eureka today is a vestigial remnant of history. The friendly community still serves as the supply point for local mines and ranches, even though the opera house is closed, the railroad tracks are gone, and the old presses are silent.

North and east of Eureka rises one of Nevada's 413 mountain ranges—the Diamond Range. Its high reaches are above tree line. It stands as an impressive north-south wall, with ranches snuggled into the heads of the valleys that wrinkle its flanks.

The Diamond Range is young, but its rocks are not. I remember walking its eastern flank in the early morning sunlight and picking up rocks. Much of the mountain is composed of fragments, large and small, of a pre-existing mountain range. The particles in these rocks were originally torn off the uplifted Antler Mountains, deposited in a downwarp to the east, and cemented back into clastic sedimentary rocks. Now, once again, they have been uplifted from the black depths into the morning light, only to be ripped apart again by erosive forces and the rock hammer of an itinerant geologist. They are Mississippian rocks.

The Mississippian was a dramatically different world from the earlier Paleozoic periods. The crushing force of arc collision uplifted a mountainous highland just oceanward from the shore. The lofty

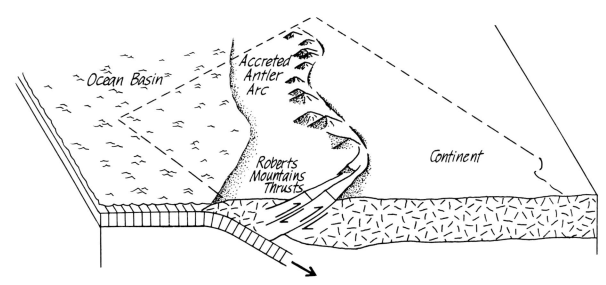

The post-Antler Great Basin.

island extended north-northeast across the entire length of the Great Basin, from California to southern Idaho. The cool rock in my hand contained fragments of marine sediment from the Antler Mountains. The great heap of stacked marine sediment created a mountain system that pressed down on the crust, and the edge of the continent foundered from the weight. A deep, ocean-filled basin formed east of the thrust front. The fragments that composed my rock washed and tumbled down into this Mississippian trough. The trough filled rapidly with the great volumes of debris that flooded down the flanks of the Antler Highlands. Deltas and turbidite fans banked in against the submerged flanks of the uplift in thick heaps of gravels and coarse angular clastics. Locally this pile of debris is more than ten thousand feet thick.

Only the finest clastics washed to the continental shoreline, far to the east of the Antler Mountains. These shales interbedded with limestones and dolomites on the shallow continental shelf. The thin Mississippian shelf sediments are only one-tenth as thick as the basin sediments.

What was the environment to the west of the Antler Mountains? We aren't certain, since there are only a few scattered outcrops of Antler-aged rocks found today to the west. These siliceous and volcanic rocks indicate deposition below the snowline in a deep-water oceanic region, so we presume it was deep ocean to the west. There is, however, a thick sequence of late Paleozoic volcanic rocks to the west of the Antler. These may not relate to Antler time, but they are perhaps part of an extensive younger island arc terrane. The volcanics are now largely buried by younger rocks.

Later events were to move the western rocks over the former Antler highland. Their present locations may not relate to their past sites of deposition. When the clues are later moved around, it is difficult to determine the original scene of the events.

We do know that the North American continent was slowly drifting northward across the equator during the late Paleozoic. In Mississippian time, the equator lay just to the south of the Great Basin, trending from Los Angeles to northern Wisconsin. In Pennsylvanian time, the Great Basin was on the equator, as it stretched from San Francisco to Salt Lake City. By the Permian, the equator extended from northern California to Minnesota and the Great Basin was just in the southern hemisphere.

TENSION?

In only twenty-five million years the original Antler Mountains were gone. This is a remarkably short time for the destruction of such a large range. The highland and the Roberts Mountains allochthon were interred by their own debris and by some later marine beds. Imaginative geologists named these overlying rocks the overlap sequence.

The Antler had at least one later pulse of renewed uplift. Geologists poking around the mountains of central and northern Nevada have found some indications of geriatric rejuvenation. Some of the original Antler rocks were uplifted and tilted in the late Paleozoic, but, once again, horizontal sediments buried the enervated rocks.

There are faint clues remaining beneath the overlying carpet that indicate that the erosive destruction of the Antler Highlands was accelerated by some violent processes. Evidence for the violence lies within the Pennsylvanian- and Permian-aged overlap beds. Large angular clastics of local origin attest to deposition within steep-walled valleys. This relief implies the possibility that the flanks of the valleys were rifted open along faults. The high relief imparted the required energy for erosion to rip large pieces of rock off the faulted valley walls. Some of the valleys dropped down so far that they fell below sea level and were invaded by the ocean. These deep troughs within the Antler Highlands may have been created by tensional forces.

Evidence of great tension or downwarping at this time also exists in the eastern Great Basin, well to the east of the Antler. Along the shelf of what is today northwestern Utah, a deep trough sagged downward during the Pennsylvanian and Permian times. This is the Oquirrh Basin. There are more than twenty thousand feet of Permo-Pennsylvanian fossiliferous limestone and generally nonfossiliferous sandstone in this basin.

Another deep valley existed in the western part of the Great Basin. Scattered outcrops near Death Valley contain turbidites that indicate deep-water deposits synchronous with the sediments of the Oquirrh Basin.

Perhaps this late Paleozoic period of tension is directly related to continental rifting. Remember, one hypothesis for the failure to find Antler island arc rocks is that the arc broke away from the continent and drifted back to the west after collision. The tensional valleys within the Antler may reflect down-dropped blocks along the edge of the fragmented continent. The Oquirrh Basin and the trough near Death Valley might be aulacogens.

SUPERCONTINENTS COLLIDE?

There are many other possible explanations. One of the most interesting hypotheses proposed for the renewed uplift of the Antler Highlands, the tensional rifting of the old uplift, and the downwarping of the Oquirrh Basin looks well beyond the borders of the Great Basin for an answer. This hypothesis relates to a collision of two megacontinents along the southern margin of what is now North America.

There were only two continents in the early part of the late Paleozoic. North America, Europe, and Asia were joined in a northern supercontinent called Laurasia. South America, Africa, Antarctica, and Australia formed the southern supercontinent Gondwanaland. In the late Paleozoic the two giant landmasses began moving toward each other. The sea floor that lay between these landmasses was subducted, and the two huge continents collided.

Africa crushed against the eastern United States. The Appalachian Mountains buckled upward as a result. Huge allochthonous sheets were thrust westward by the collision. South America closed against our southern border, thrusting an allochthon of oceanic sediments from south to north across the southern continental shelf. These rocks are exposed in Oklahoma's Ouachita Mountains. This colossal impact occurred in Pennsylvanian time but was not completed until Early Permian. All the continental masses on earth were now joined in one megacontinent called Pangea.

The timing is right. Perhaps the crushing closure of an ocean basin and the pressures of supercontinent collision were sufficient to nudge the North

Toquima Range, Nevada. Compression forces rocks on top of each other. Much of the history of the Great Basin has involved massive episodes of pressure from the forces generated by the jostling of crustal plates. *Bill Fiero.*

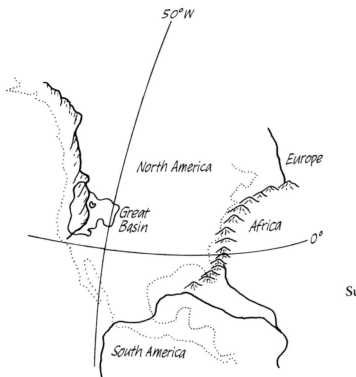

Supercontinents collide (from Coney).

American Plate with enough force to create chaos well to the west of Virginia or Oklahoma. The ancestral Rockies uplifted in Colorado and Utah east of the Great Basin at this time. This distant event may have caused the tension in the Antler Mountains, downwarped the Oquirrh Basin, and dropped the trough near Death Valley. In human affairs, local events are often the result of distant forces. As we learn more about the earth, we are increasingly discovering that the same reality often applies.

THE ARRIVAL OF SONOMIA

The Great Basin region had barely recovered from the collisional event of the Antler island arc crushing into the western margin of the continent when it rifted away and a new ocean basin formed. Parts of the region were downbuckled by great tensional forces generated either by the rifting away of the arc

or from continental collision well to the east and south.

For more than twenty years, geologists have recognized a major tectonic event during the Late Permian and Early Triassic in the Great Basin that is comparable in magnitude to the middle Paleozoic Antler orogeny. The ideas that may explain this mountain-building event are just beginning to come into focus.

Once again, just as in the earlier Antler orogeny, the evidence shows deep-water sediments of the offshore environment to be thrust over the eastern shallow-water rocks. The late Paleozoic turbidites, siliceous clastics, and volcanic rocks that had accumulated in the post-Antler western ocean basin were forced over the rocks of the eroded Antler Highland Belt of the same age. The western rocks came to rest on both the Antler allochthon and rocks of the eastern assemblage. These are the Golconda thrusts, and this is the Sonoma orogeny. There was easterly movement of at least forty to fifty miles. The effects of the Sonoma orogeny are recognized from north-central Nevada south and east to California.

The Sonoma orogeny, like the Antler, is an atypical orogeny. There is no evident metamorphism, intrusion, or deformation of the underlying rocks. The deformation is entirely within the thin-skinned Golconda thrust sheets.

These are the facts. They are what geologists have pieced together from years of climbing through the dry hills of the Great Basin. But they don't tell us how it happened. What caused the Sonoma orogeny? Why is it, like the Antler, so dissimilar from other mountain-building episodes? How did this accumulation of deep oceanic rocks get displaced eastward in thin sheets for such a great distance over the shallow-water shelf deposits? The fundamental questions are left unanswered. Like the three-year-old child who keeps asking "Why?" in response to each of our answers, geologists also keep asking "Why?" First we piece together the evidence; then we need the hypotheses to match the data.

There are many proposals to choose from, since geologists love to speculate on thin shreds of evidence. Today, most models involve plate tectonic ideas for an explanation. None of the concepts are proven, and perhaps they are incapable of proof. One of the most recent ideas, however, presents an interesting scenario.

A collision occurred between the North American continent and another wandering stranger. In late Paleozoic time, the western continental margin of North America was essentially the same as that established by the rifting in the Late Prepaleozoic. Of course, this margin was modified by the arc collision of the Antler in the Late Devonian and by additional rifting in the late Paleozoic. At the close of the Paleozoic, this borderland was the site of yet another collision. A microplate called Sonomia migrated from a distant but unknown site toward the continental margin. The sea floor adjacent to the North American continent was firmly fixed to the continent. The approaching Sonomia rode over the top of the sea floor. As the sea floor subducted below the microplate, it melted and the lightweight fractions rose upward into Sonomia, creating island arc volcanics. Punctured with these volcanos, the smoking microplate moved against the continent.

The sediments that today compose the Golconda thrusts had been deposited on the sea floor adjacent to both Sonomia and the continent. The source of those sediments was probably both regions. The impending collision bulldozed the sediments off the consuming sea floor and squeezed them between the two crustal pieces.

Eventually, the oceanic crust was completely consumed and Sonomia and the continent collided. Sonomia probably first collided with the continent in its northern reaches and then rotated as it moved southerly against the northwest-trending continental margin. The collision occurred in the time range between earliest Triassic and the end of the Middle Triassic. The continental margin, including the shelf sediments, initially descended westward along the subduction zone, forcing itself under the siliceous and volcanic rocks of the approaching Golconda sedimentary wedge. The crust, however, was rich in silica and was lightweight. It was unable to descend substantially below the microplate, and westward subduction ceased.

The Golconda thrusts can be interpreted as the slide zone along which the subducting plate slid beneath the Golconda sediments. Some of the folds and thrusts within the allochthon formed during growth of the island arc before plate collision. These structures moved along with the microplate and were probably transported a long distance, which may explain why the rocks below the allochthon are so little affected by the collision. It also solves the problem of how to thrust a thin succession of sea floor sediments for forty to fifty miles. At the time of final underthrusting, the Golconda sediments were probably a very thick mound of folded and faulted rocks flanking Sonomia.

As the pieces collided, the Golconda sediments and the underlying lightweight continental margin were lifted upward. Sonomia and the continent were

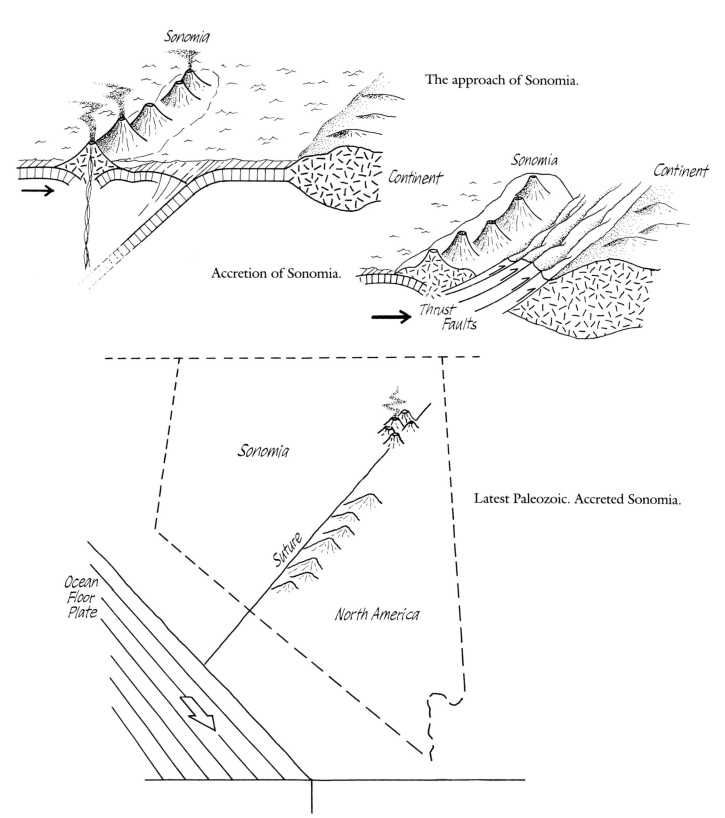

Sonomia

The approach of Sonomia.

Sonomia

Continent

Continent

Accretion of Sonomia.

Sonomia

Thrust
Faults

Sonomia

Suture

Latest Paleozoic. Accreted Sonomia.

Ocean
Floor
Plate

North America

now firmly sutured. The ocean floor continued to move against the sutured microplate and the continent. Now the subduction was east-dipping and took place along the western or outboard side of Sonomia. The subducted sea floor now lay beneath Sonomia and the continent. The rising recycled melted fraction of the ocean floor rose as intrusives and volcanics along a zone about 60 to 120 miles inland from the western edge of Sonomia. This new magmatic arc was built during Late Triassic time. This time the volcanos were present on the continent rather than in an offshore island arc.

Sonomia added significant local girth to the continent. The northwestern one-third of Nevada and almost one-half of northern California might be exotic terrane that was pieced onto North America during this collision.

But geologists are argumentative folks. In true scientific style, they challenge hypotheses. Perhaps a collisional model with one microplate, Sonomia, is too simple. Maybe several discrete microplates accreted to the North American margin. These smaller chunks would comprise the region called Sonomia. The discussion will continue until conclusive field and lab evidence is found and agreed upon.

Whether there is one microplate or several, microplate accretion is a model to explain the Golconda thrusts and the Sonoma orogeny. Sonomia would be the first and oldest of a plethora of terranes that are interpreted to have accreted to western North America in more recent times. The western part of the continent has grown as a result of this gradual piecing together of tectonic plates, much in the fashion of a collage. Alaska, British Columbia, most of Washington, Oregon, and the remainder of California were added to our continent by the collision of exotic strangers from out of the sea. The story of western North America is apparently a tale of the building of a collage from previously shattered pieces of our continent and remnants of other continents. It is a fascinating story, but the evidence lies to the west of the Great Basin, so it is not a tale to be told here.

SONOMIA ROCKS

What can we decipher of the past of our newly arrived neighbor, who has forced himself into our midst? What portfolio does the ambassador carry? After all, the rocks of about one-third of Nevada, including the basement for Reno and a large part of northeastern California, are apparently exotics.

These are difficult questions to answer, since most of Sonomia is buried by Mesozoic or Cenozoic rocks. Windows through the cover expose Permian or perhaps Early Triassic volcanic rocks. They occur largely as fractured or fragmented rocks and also as lava and intrusive masses. Associated turbidites and other clastic sedimentary rocks such as conglomerates composed of pieces of volcanic rocks appear to have been deposited on a marine shelf or oceanic basin. There are few if any rocks in Sonomia that are of continental origin. Rocks as old as Ordovician have been identified. Arc volcanism was apparently present from Ordovician to at least Permian, indicating a long history of association with subduction zones. These are rocks with a checkered past.

A great thickness of rhyolites forms the base of the overlying strata. These Early Triassic ash-flow tuffs were deposited on a block-faulted terrain. Perhaps they represent the generation of magmas in Sonomia after it collided with the continent and the microplate was stretched by the underriding continental margin.

Where did Sonomia come from? Again, we don't know. It has been suggested that Sonomia is the return of the island arc that collided with North America in the Devonian and was responsible for the Antler orogeny. In this model, after the Antler collision the arc moved seaward in late Paleozoic and remained offshore. A marine basin on the eastern side of the arc captured the sediments of the Golconda allochthon. When the arc returned for a second collision, it accreted to the continent, causing the Sonoma orogeny. This hypothesis creates an interesting and perfectly logical explanation. Unfortunately, try as hard as we can, we geologists have been unable to find undisputed Antler arc sediments in Sonomia rocks. To date, therefore, there is no conclusive field evidence to support the idea that the same arc is responsible for both the Antler and the Sonoma orogenies.

If Sonomia's genealogy is not related to the Antler, what is its origin? Let's frame an alternate hypothesis. Paleontologists have noted that marine creatures of Sonomia sediments resemble North American faunas more than those of other continents. Maybe Sonomia did not make a transoceanic migration. Perhaps Sonomia is a wayward piece of North America. It may be a prodigal son that took a journey only to return a much-changed sojourner. It could be a piece from the north coast. Perhaps it is a chunk of British Columbia that separated from the continent and drifted to the west during the late Paleozoic. The island arc and Golconda sediments were formed when the fragment reversed course and headed back to its North American birthplace in the latest Paleozoic. We can only hope that the Canadians don't lay claim to what might be a former piece of their territory, or all the political maps of North America will require monumental changes.

If the Canadians should push their claim, we in the United States might have our own. A slice of Nevada-California might now be found securely lodged in Mexico. If you take a map of the trends of the Paleozoic sediments and mountain belts in western North America, you will quickly see that the trends are at right angles to the later Mesozoic trends of central California. Perhaps the projection of these Paleozoic trends was sheared off before the arrival of Sonomia. The truncated edge of the old continent would not have been large, and perhaps it will yet be found within the complex Mesozoic terranes of California. However, there are some rocks in northwest Sonora, Mexico, that bear a suspicious resemblance to the Great Basin Paleozoic trends. To date, it is not yet established that these rocks are the missing chunk, so we will refrain from asking for a decision by an international court of land claims. When you observe the impunity with which nature shuffles and recombines lands through geologic time, the paltry claims by one political body or another over a piece of territory seem insignificant. Nature, moreover, has plenty of time to exercise territorial sovereignty.

11 The Mesozoic

THE ROCK-BUSTER: THE CORDILLERAN OROGENY

Throughout the 4.5-billion-year history of the Great Basin there have been many significant geologic events. None, however, was more consequential than the Cordilleran orogeny. This mountain-building episode radically transformed the western one-third of the continent. Great tectonic movements swept in waves across our region from Late Jurassic to the Eocene.

So immense were these tectonic movements that geologists originally described them as being a series of orogenies. Current plate tectonic concepts more logically relate the mountain building to one almost continuous period of tectonism. Diastrophism on this scale could only result from the motions of major plates along and under the western margin of North America. It is convenient for clarity of description, however, to divide the tectonism into three distinct phases. The earliest phase, the Nevadan, describes the Middle to Late Jurassic movements. This rock deformation and time of plutonic emplacement largely involved the western Great Basin. By Late Cretaceous time, the Sevier phase was thrusting rocks along the eastern margin of the Great Basin. The orogeny continued through the Cenozoic Eocene, when the Laramide phase was deforming the Rocky Mountain region. North America, from California to Colorado, from Canada to Mexico, was reshaped by this great rock-buster—the Cordilleran orogeny.

THE QUIET BEFORE THE STORM: THE TRIASSIC

The early Paleozoic history of the Great Basin is one of gentle subsidence. In the Devonian, the Antler Mountains were uplifted and soon eroded. Again, 235 million years ago, near the close of the Paleozoic, orogeny uplifted the central Great Basin. This time the uplift was the result of the collision of the microplate Sonomia with the continental margin. The late Paleozoic arc welded firmly to the continent. As Sonomia and the continent ground together, the intervening oceanic floor was completely consumed by subduction. A new subduction zone began, far to the west on the western borderland of Sonomia. This time, easterly subduction slid the ocean floor under the sutured Sonomia. The collision of Sonomia uplifted the continental margin, and it remained a highland in the central Great Basin throughout most or all of Triassic time.

Basins formed on either side of the uplift. The history of each of these basins is radically different through early and middle Mesozoic time.

The Eastern Basin

The eastern side of the upland was flooded by a sea. Fine clastics and limestones were deposited in the warm shallow waters. Rivers carried muds and other clastics off the continent to the east. Some fine clastics washed down the flank of the western highlands and built broad deltas. Floods washed trees and plant debris into the basin and their fragments remain today, imprinted in the rocks.

In the Late Triassic, the region to the east of the uplift slowly emerged from the sea. Sandstone, conglomerate, and claystones, all containing abundant volcanic detritus, were deposited on the floodplains of this continental area. The windblown volcanics settled to the ground and were transported northerly by streams.

The major source of clastics for the eastern basin was the highlands to the east and south in Colorado and New Mexico. These latest Triassic rocks are dominated by red beds. They add an exciting splash of color to the mountains of the southern and eastern Great Basin today.

The Western Basin

The history of the western basin is strikingly different. The Sonomia microplate had been heated while en route to its continental collision by the upward intrusion of subduction-melted magmas. Volcanos crowned the higher parts of the microplate. After collision 235 million years ago, the subduction zone migrated farther to the west and Sonomia igneous activity quieted down and eventually ceased. As the sutured plate cooled, it contracted and began to subside, forming a deep basin to the west of the central Great Basin highlands. The eastern and southern hinge lines of the sinking region are approximately coincident with the suture zone between the continent and Sonomia.

When the microplate was in the early phase of subsidence, 235 to 215 million years ago, the deepwater offshore basin was the site of fine carbonate and siliceous mud deposition over the late Paleozoic basement. Pebbles of Paleozoic debris were carried down the flanks of the uplift by turbid clouds of sediment. They settled into the depths.

To the southwest of the now firmly welded, defunct, and subsided Sonomia was an ocean floor plate in contact with the western border of the North American continent. This oceanic plate is thought to have been moving southeasterly in a lateral motion along a major transform fault. Strontium isotope data from more recent granitic intrusions show a lack of continental rock to the west of this transform. Consequently, some geologists postulate that this faulting detached a fragment of the continent. They are presently searching for this fragment in Mexico.

THE STAGE IS SET

Just before the end of the Triassic, about 220 to 215 million years ago, a subtle change began to take place in western North America. Hardly noticeable at the time, this change would result in the greatest deformation of the western part of the continent since the earliest Prepaleozoic times. As the Chinese sages said, the longest journey begins with the first step.

North America, and its now-accreted Sonomia, changed its direction of motion relative to the western ocean floor plate. The continent ponderously began to rotate clockwise. The motion forced the oceanic plate to obliquely subduct beneath the continent. The lateral motion of transform faulting was now out of vogue. The new continental margin tectonics for the remainder of Mesozoic and most of Cenozoic time would reflect the latest style of deformation: eastward directed head-on collision between sea floor and continent. Subduction beneath the continent is the latest thing, with the oceanic Kula Plate now moving under the continental margin and therefore under the Great Basin. This subduction marks the beginning of the modern circum-Pacific orogenic system.

Responding to the onset of this new style, volcanism began along the continental margin as the subducting plate dove into the subcrust and melted. The Triassic western basin was flooded with both volcanics and sediments derived from the erosion of these continental volcanos. The basin bulged upward from the thermal expansion related to volcanism. What had been a deep trough became a shallow marine depression and remained so through most of the middle part of the Late Triassic. The southern portion of the basin flexed upward so high it rose above sea level. This uplift extended at least as far south as today's Death Valley. The emergent land mass rose approximately along the Sonomia suture line. The earth has a long thermal memory of former events. First a suturing, then a contraction and subsidence, and now a thermal expansion—all along approximately the same line. Where there was once activity there is often action again.

An enormous volume of finely ground-up clastics, derived from the Triassic volcanics, flooded the still existent northern basin. Rivers, flowing north to the eastern margin of the basin, transported these

Sierra Nevada. Intrusions from the subducting Mesozoic sea floor have been uplifted during later geologic movements. These hard igneous rocks resist the erosion of water and ice and stand as jagged peaks along the western edge of the Great Basin. *David Slemmons*

sediments to the ocean. What a beautiful Nevada shoreline this must have been. White quartz sand beaches rimmed the basin. In a few places rivers built huge deltas that flooded the beaches with muds. Turbidite currents carried the sediments from the deltas and shores to the floor of the deep basin.

About two hundred million years ago, near the close of the Triassic, volcanism waned and may have ceased. Without a heat source to maintain the thermal bulge, the southern volcanic mound subsided. It slowly sank below sea level again and was covered partly or continuously by carbonate banks. Few clastics now reached the basin. The stage was set for the greatest period of igneous activity yet to be seen in the Great Basin.

PANGEA BREAKS UP: THE JURASSIC

An earth-shattering historic event occurred near the onset of the Jurassic. Pangea began to break up. This profound change resulted in global-scale tectonics. Earth would never be the same again.

A thermal welt began to grow in the early Mesozoic along the old suture zone between North America–Europe and South America–Africa. As the long east-west bulge rose, it cracked and the continent began to come apart at the seams. Slowly the megacontinent split along a series of triple junctions. An arm of the sea flowed into the rupture. The failed arms of the triple junctions became the sites of some of today's major rivers—the Niger, Amazon, Connecticut, and Mississippi. The huge continental masses pulled away from each other in ponderous slow motion, and the North American–European continent moved off in a northwesterly direction. A youthful ocean now separated the land masses.

The breakup of Pangea accelerated North America and caused it to slowly rotate clockwise relative to the sea floor plate to the west. Far to the west in the Great Basin, this change of direction resulted in the head-on collision of the sea floor with the continental margin. The agonies of the birth of an eastern ocean would be felt all the way to the Great Basin. Atlantic motions would continue to profoundly influence the far west.

The effects of this rifting were widespread throughout the newly liberated North American continent. The continental interior was isolated from the newly formed and growing Gulf of Mexico by a thermal bulge where the continent had rifted. The rivers that formerly flowed southward were blocked. The highlands in the central Great Basin, uplifted by Sonomia collision, blocked many of the drainages to the west. The sediments eroded from the eastern interior were impounded. They were trapped in a large basin that occupied the eastern Great Basin, Colorado Plateau, eastern New Mexico, and part of Texas. The basin slowly subsided, perhaps in response to the weight of the sediments. All but the southwestern Great Basin remained just above sea level. The captured sediments within the isolated basin rusted in the hot Mesozoic sun until they became a bright orange or deep red-brown. Some of the most beautiful scenery in the Southwest, such as the Painted Desert, is developed by the present-day deep erosion of these multihued early Mesozoic shales.

During the Jurassic, sands were blown generally southward through the red bed basin. Heaped by the winds into high dunes, the sandy desert covered the eastern Great Basin. Some of the sands even blew to the west where they have been found within the Jurassic arc volcanics in the southwestern Great Basin. These fossilized dunes today are the brightly colored red and white cliffs and rounded hummocks that delight visitors to such well-known areas as Valley of Fire, Red Rock Canyon, and Zion National Park. By the end of the Early Jurassic, the eastern Great Basin was uplifted. Sedimentation ceased and the land was exposed to severe erosion.

The Sonomia upland in the central Great Basin may have been eroded down to or below sea level by Jurassic time. Geologists find little or no evidence of nearshore deposits in the western oceanic trough at this time. The sediments are only indicative of the open sea, supporting the idea that the highland created by Sonomia was eventually eroded

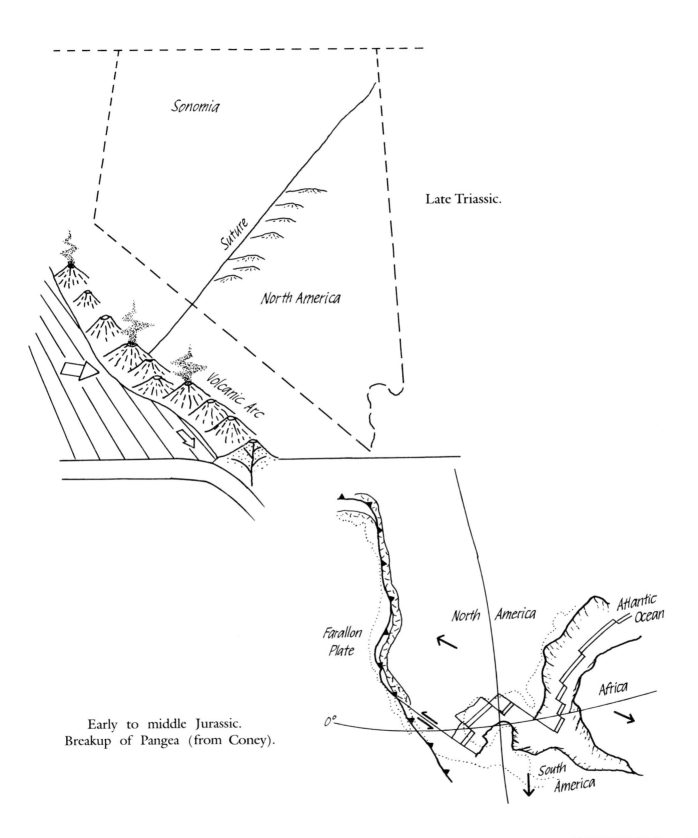

Sonomia

Suture

North America

Volcanic Arc

Late Triassic.

Early to middle Jurassic.
Breakup of Pangea (from Coney).

Farallon
Plate

North America

Atlantic
Ocean

Africa

0°

South
America

away and the remnants slowly foundered below the sea.

As the Gulf of Mexico widened in response to the separation of South America from North America, Yucatan pulled away from Texas. A large tear, the Mojave-Sonora Megashear, is interpreted by some geologists to have ripped across Mexico from Los Angeles to the Gulf, displacing to the southeast all the pre-existing terranes to the south of the megashear.

THE NEVADAN PHASE OF THE CORDILLERAN OROGENY

The Batholith

Batholith—deep stone. The name leaves one with the impression of a massive lump of deeply buried rock immersed in the realm of Pluto. In fact, the Sierra batholith is a composite mass composed of hundreds of distinct plutons. Some of the intrusions are as large as one thousand square miles. The pluton is almost seventy-five miles wide and it extends almost the entire length of California.

The batholith is now raised from the stygian depths. Today, this is the Range of Light. John Muir so described his favorite mountain haunts, the Sierras. Pale gray granodiorites stand in jagged spires and rounded domes above glacially scoured azure lakes. A wall of rock stands along the western edge of the Great Basin, effectively blocking both the early emigrants to California and land-bound travelers today. Why so much granite? Earlier orogenies were essentially devoid of intrusives. What is the source of the granite rock? Why so many intrusives concentrated within this range at this time? Why the Sierra?

Subduction

The earlier orogenies, Antler and Sonoma, were arc or microplate collisions with subduction zones dipping westerly under the impinging sea floor. The resulting volcanos pierced upward through the offshore plate. These rising magma plumes were influenced only by the original chemistry of the subcrustal rocks and by the ocean floor plates through which they intruded. These were oceanic volcanos.

The Mesozoic is the time of the Nevadan phase of the Cordilleran orogeny. With the onset of the convergence of the Kula oceanic plate against the North American Plate in Late Triassic, and particularly in the direct convergence of Jurassic-Cretaceous time, the subduction zone had a different polarity from the earlier orogenic events. This time the sea floor subducted easterly beneath the continent. Earthquakes shuddered the continental margin as the great slab of oceanic lithosphere rubbed under the continent.

The subducting plate melted as it plunged to the depths below the continent. The lightweight oceanic sediments, traveling piggyback down with the subducting slab, melted. As these former sediments melted, the light elements separated from the heavy. The low-density distillations were lighter than the surrounding deep rocks. They rose as great tear drop—shaped masses. The rising silica-rich magmas were further contaminated by the silicic chemistry of the deep crustal rocks through which they intruded, such as old sedimentary, volcanic, or granitic rocks. The ascending molten blobs were thus heavily influenced by continental chemistry, and the composition of the magmas was altered. These high-silica rocks have the right chemistry, when melted and then cooled slowly, to make granites. Hence, the granites of the Range of Light.

The plutons rose almost vertically as hot, mobile masses. As the magmas elbowed their way into the crust, they needed room. They shoved the sediments aside while simultaneously baking and pressure-cooking them into metamorphic rocks. Careful observation of the rock faces of the Sierra reveals easily seen internal flow banding of the originally molten material. The large compressive forces resulting from this intrusion squeezed the rocks of eastern California and piled them up into thrust sheets.

The Timing

Intrusion first began in the western part of the Great Basin during the Late Triassic. By the middle

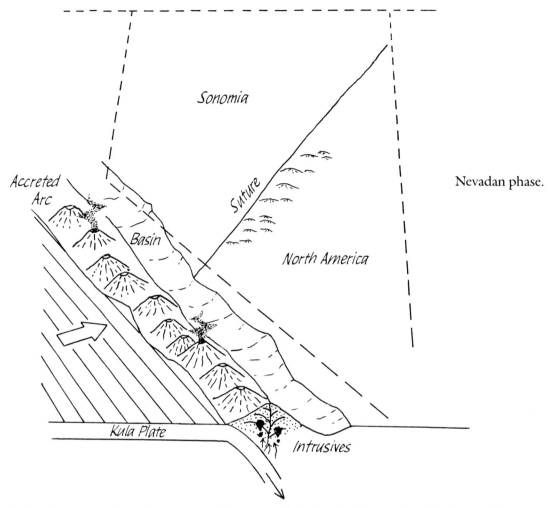

Nevadan phase.

part of the Early Jurassic, plutonism was widespread. The greatest time of intrusion was during the Late Jurassic to Early Cretaceous. The eastern half of the batholith is mostly of Late Cretaceous age, according to potassium-argon radioactive age determinations.

There are many smaller intrusions east of the Sierra Nevada. They extend across eastern California and Nevada to the Idaho batholith. Potassium-argon radioactive age determinations relate these easterly plutons to Late Jurassic and Cretaceous times.

A great mountain range resulting from intrusion and compression began to emerge in Late Jurassic time. Calculations indicate that tens of thousands of feet of rock were removed by erosion from the roof of the batholith during this time of Mesozoic uplift, exposing the underlying intrusive granitic rocks to erosion.

The rock of the Sierras is Nevadan. The great batholithic intrusions of the western United States are the most impressive geologic consequence of this phase of the Cordilleran orogeny. The Mesozoic is thus one of the greatest periods of granite formation in all of earth's history.

From What Depth?

Geologists have discovered an interesting relationship in the chemistry of igneous rocks. The ratio of potassium to silicon increases with the increasing depth from which magmas originate. The granitic

Sierra Nevada. The light-colored Mesozoic intrusions stand above Convict Lake. Dark-colored metamorphics, trapped by the rising magmas of the intrusions, are frozen within the rocks. *Bill Fiero*.

Valley of Fire. Mesozoic sandstones are stained red by iron oxides. These dramatic color contrasts characterize the Mesozoic. *Bill Fiero.*

rocks of the Sierras show a systematic eastward increase in their potassium to silicon ratio. This is precisely what would be expected in intrusives related to the melting of an inclined, subducting slab of oceanic lithosphere as it plunged ever deeper in an easterly direction beneath the continent.

The protomagmas were probably generated in or near the subducting slab more than seventy-five miles down, but the final magmas must have incorporated much material from the crust through which they passed on their rise to the surface. According to the potassium-argon clues in the rock, the rising plutons finally reached an equilibrium with the surrounding crustal materials at a depth of only two to five miles. Here the molten masses congealed into rock.

Volcanics As Well

Some of the granitic magmas, generated by the melting of the Kula Plate, were able to breach the continental crust all the way to the surface. Volcanos erupted silicic magma in enclaves between the granitic plutons. Voluminous quantities of ash were windblown from the volcanos and are preserved within Mesozoic sedimentary rocks to the east. Although many of these are undated, most range from at least Late Triassic to Middle Cretaceous and are the extrusive counterparts of the Sierra plutons.

With such extensive extrusive activity, why is there so much ash and so little lava? The answer again lies in the nature of the rock chemistry. The rising magmas become saturated with silica through contamination by the continental crust. Silicic magmas are sticky and viscous. They lack the fluid nature required for lavas and instead are highly explosive. The magmas erupt at the surface in shattering blasts, creating great amounts of ash.

In the eastern Great Basin, many Late Triassic and Middle and Late Jurassic shales are rich in silicic ash. The Late Jurassic rocks are widespread in Utah as multihued clay-rich shales. The most voluminous ash in the shales of the western United States spans the interval from the middle of the Early Cretaceous to Late Cretaceous, or from about 120 to 70 million

years in age. This is also the time span of greatest silicic magmatism in the Sierra region of the western Great Basin. It has been estimated that there are perhaps more than one million cubic miles of altered ash in these shales. The correlation of the ash strata of the eastern Great Basin to the known voluminous silicic magmatism farther west in the Great Basin is impressive.

Regional Trends

The granitic batholiths and their surface-erupting cousins, the volcanics, form a large-scale belt that is virtually continuous along the entire western margin of the Great Basin. They are part of a regional trend that extends from Canada to Mexico. The plutons stand on Prepaleozoic crust to the south, but in the Great Basin they are partially built upon the exotic Sonomia terrane inherited from the Paleozoic.

The rocks intruded by the eastern part of the Sierra batholith are thoroughly metamorphosed. They may be the extension of the Paleozoic depositional and tectonic belts of the Great Basin. The shelf, transitional, and deep-water sediments, as well as the rocks disrupted by the Antler collisional event, should extend into this western region. Some geologists believe they can recognize the metamorphic equivalents of these rocks. Some even identify the Antler event by the more intense deformation of early Paleozoic rocks in that part of the Sierra into which the Antler terrain should project.

The Big Picture

From the Late Jurassic to the Late Cretaceous, the combined North American–European continent continued moving away from Africa, widening the intervening ocean. About 135 million years ago, South America also separated from Africa and the Atlantic Ocean began to tear open from south to north. The Kula sea floor plate, lying to the west of North America, continued to subduct at about three inches per year under the Cordilleran region.

The resulting Mesozoic tectonic pattern in the western United States is therefore clearer than the pattern of earlier times. The western Cordillera ma-

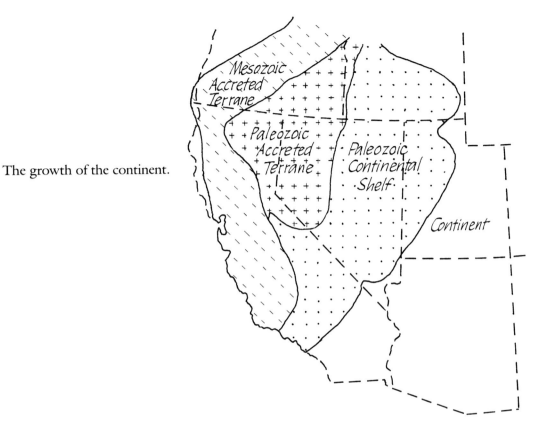

The growth of the continent.

tured into a tectonic belt similar to the belt found along the western side of South America today. This Andean-type orogenic belt created a volcanic and heavily intruded mountain chain above the east-dipping subduction zone.

Suspect Terranes

Geologists who have worked along the west coast of North America have found some intriguing rocks that did not seem to relate to the surrounding North American geology, but their origin was unknown. They were so exotic that they were called suspect terranes. We suspected they were not natives, but we didn't know they were aliens, disguised as residents, lodged into the continent.

Now many geologists believe these peculiar terranes were scraped off sea floor plates as they subducted under the continent throughout most of Mesozoic and Cenozoic times. Anything that was resident on these sea floor plates was carried along

as though on a gigantic conveyor belt. Islands, oceanic rises, marine sediment, shark teeth—everything was crushed up against the edge of the continent. Bulky oceanic island arcs and far-traveled microcontinents were also accreted to the continent as a collage of exotic blocks. Strange rock assemblages, termed ophiolites, resulted when island arc, oceanic plate, and marine sediments were all smashed together in the pile-driver blows of exotic terranes crunching into the continent. Ophiolites are often found along the suture lines that separate these suspect terranes from the native rocks. Calling them suspect terranes begs the question as to how far they traveled before sticking to the continent. Some paleomagnetic memories locked within the grains of these foreign rocks suggest large north-south travel distances.

After each collision, the subduction zone would step westward to the outboard margin of the newly accreted blocks. New pieces were continually being

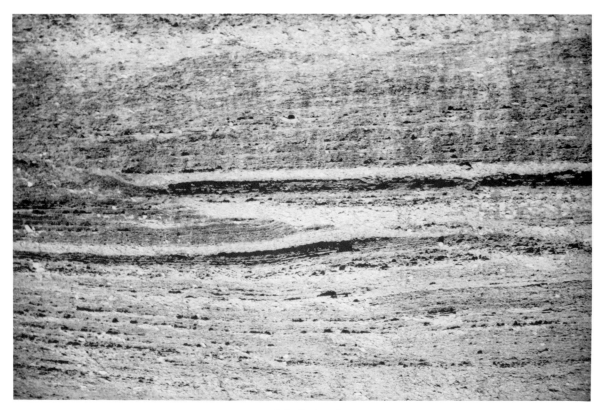

Schell Creek Range, Nevada. Compressive forces during the Sevier phase have thrust rocks together, forcing some to rise over their counterparts. Such compression occurred during several periods of the Paleozoic as plates collided with our continent. *Bill Fiero.*

coupled on the oceanward side of the new continental border. Alaska, British Columbia, most of western California, and portions of Oregon and Washington were constructed in this manner. Californians are a diverse lot of people, and so is their land.

THE SEVIER PHASE OF THE CORDILLERAN OROGENY

The eastern Great Basin was essentially undeformed from the Prepaleozoic until the Early Jurassic, except for the downwarping of the Oquirrh Basin in the late Paleozoic. The Antler and Sonomia collisions, which heavily affected the central and western portions of the region, did not influence the slow

but steady horizontal deposition of sedimentary rocks to the east. The great intrusions of the Nevadan phase were also far to the west. This time of quiet was brought to an abrupt close in the Middle Jurassic.

An Accelerating Plate

The western margin of the continent during the Mesozoic was dominated by tectonics related to the subduction of the Kula Plate beneath the North American Plate. A lofty mountain chain of Andean character rose above the melting sea floor plate. Allochthonous terranes drifted on the Kula or Farallon sea floor plates and accreted to the continent's western edge until middle Cretaceous time, when the Cordilleran tectonic collage was completed. North America grew by about 30 percent through this tectonic accretionary addition in the western Cordillera. Eastward subduction of oceanic lithosphere was continuous during this time interval. East of the arc mountains, the sediments eroded from the highlands were shed into a basin that was generally continuous from Canada to southern Nevada.

Compression is the result of such convergence of plates. Compression also results from the forceful intrusion of plutons into the crust, such as the plutonism of the Nevadan phase. Compressive forces thus began in the Jurassic and continued throughout the Cretaceous. Another factor also accounted for the eastward compression. The North American Plate began to accelerate. As it increased its velocity relative to the western Farallon Plate, the oceanic slab began to slide under at an increasingly lower angle. Consequently, it would directly interact with the overlying continental plate for a great distance inland. This placed additional compressive forces on the overlying continental crust.

The Overthrust Belt

Compression ruptured the continent well to the east of the continental borderlands. Huge sheets of sediment were shoved easterly along thrust faults. Tens of trillions of tons of rock were shoved over other rocks. Whole mountains were forced over each other. These great, almost horizontal faults ripped the continent along a rough north-south line through the eastern portion of the Great Basin in what is termed the Sevier phase of the Cordilleran orogeny. The great band of thrusting resulting from the Sevier compression is termed the Overthrust Belt.

The dominant rock types involved in the movement are the continental shelf strata of late Prepaleozoic and early Paleozoic age. This great wedge of easterly thinning sedimentary rocks was convoluted by the compressive forces. Some of the beds were folded, but most were broken by the great thrust faults. Many of these faults were so horizontal that they slid the rocks along the original bed-

ding planes. The particular style of deformation was controlled by the character of the local rocks. Very thick, resistant units produced great overthrust sheets. Alternating hard and soft formations produced less closely spaced parallel thrusts. Soft rocks were deformed by crushing together. Sliding the rocks together, similar to stacking a deck of cards, created a thick pile. Local troughs formed in the pile during the tectonism and coarse rocks accumulated in them. These strata also included carbonate and gypsum that may record deposition before the region was heavily deformed and uplifted.

This Overthrust Belt extends essentially unbroken from Wyoming to southern Nevada, and the northernmost reaches may be traced to Canada. The belt is injected by younger plutons to the southwest, which helps to date the thrusts. It is eventually truncated by the San Andreas Fault System. The allochthonous rocks were transported east-northeastward a distance of probably more than sixty miles.

The thrusting began in the Jurassic but was primarily a Cretaceous phenomenon. The front of the sliding pile moved eastward with time. Major thrusting ended soon after the time that the Laramide phase began in the Late Cretaceous.

This compression and thrust motion warped the Great Basin. The eastern portion was differentially pushed to the southeast and the western side moved to the northwest. This right-lateral drag bent the Paleozoic northeast-southwest trends of the shelf, Antler thrusts, and the suture zone of Sonomia. These bends remain obvious on geologic maps today.

The Sevier phase is particularly well represented in southern Nevada by a series of eastward-directed folds and overthrusts. These carry late Prepaleozoic and early Paleozoic sedimentary rocks eastward over late Paleozoic and early Mesozoic rocks. These compressionally deformed rocks lie in a belt sixty miles wide. These thrusts may flatten with depth. Tectonic shortening caused by thrusting in southern Nevada is at least twenty-two to forty-five miles, although some have suggested as much as eighty

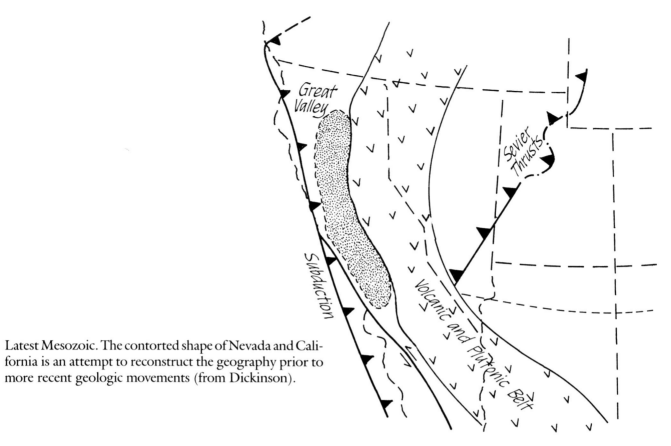

Latest Mesozoic. The contorted shape of Nevada and California is an attempt to reconstruct the geography prior to more recent geologic movements (from Dickinson).

miles. Late Cretaceous thrusting is indicated in these regions, although some deformation may be Tertiary. Farther west, in the Keystone Thrust Fault, a radiometrically dated pluton of middle Cretaceous age or older cuts the fault, substantiating older ages in the west portion of the Overthrust Belt. Some of the Sevier thrust faults are intruded in eastern California by Late Cretaceous plutons, indicating an end to at least part of the Sevier phase in the western Great Basin at that time.

Post-Sevier Basin

The weight of the advancing, thickening wedge of thrusts downbowed the crust, producing an asymmetric basin to the east of the thrusts. Clastic rocks were eroded off the thrust wedge and covered the deeper western portion of the downbowed basin with conglomerates and deltaic deposits. As the thrusts moved eastward, the axis of the basin also moved to the east.

By the Late Cretaceous, the eastern boundary of the downwarped basin lay close to the eastern boundary of the Great Basin, trending roughly northeasterly through southwest and central Utah. The basin associated with Sevier thrusting was so depressed by the weight of the thrust pile that a marine invasion flooded the basin from the south during part of the Jurassic and Cretaceous.

THE LARAMIDE PHASE OF THE CORDILLERAN OROGENY

The simple volcanic-plutonic belt of Mesozoic time began to break up about eighty million years ago in the Late Cretaceous. First the Sevier thrusts deformed the eastern and southern Great Basin. Then the Laramide phase of the Cordilleran orogeny affected the continent six hundred miles from the plate margins and far to the east of the Great Basin Sevier Overthrust Belt. Why?

Major changes in deformational style, an easterly

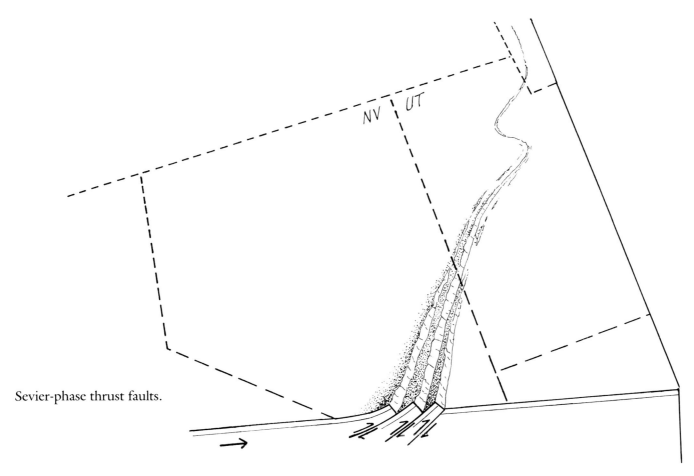

Sevier-phase thrust faults.

wave of volcanism, and mountain building far to the east of earlier tectonism indicate a major plate tectonic cause. Part of the answer is found beneath the cold waves of the North Atlantic Ocean. The magnetic stripes trapped within the igneous rocks of that old sea floor indicate there was an increase in the spreading rate along the mid-Atlantic Ridge. We don't know why.

Remnant magnetism on the floor of the Pacific Ocean indicates that it was moving northward at a rate of about six inches per year between eighty and forty million years ago. These high-speed motions of both the North American and Pacific sea floor plates increased the impact velocity between the westerly oceanic plates and the North American continent beginning about eighty million years ago. This acceleration would have caused a flattening of the angle by which the sea floor plate subducted be-

low the continent. A lower angle of subduction would put the young, hot, oceanic Farallon Plate in direct contact with the overriding continental slab. The easterly compression and the migration of volcanic activity to the east during the Laramide is thought to be the direct result of the increasingly lower angle of impact. The easterly shift of activity was also undoubtedly compounded by the easterly movement of the downward bend of the subducting plate.

The inclination of the subducting slab approached horizontal when the plates accelerated to maximum velocity. Arc magmatism died. The slab did not penetrate sufficiently into the depths of the earth to generate magma. This move to the east signaled the beginning of the Laramide phase of the Cordilleran orogeny. During the height of Laramide compression when the plates were colliding with maximum

Newberry Mountains, southern Nevada. A granite dike
cuts across an older granitic intrusion. *Bill Fiero*.

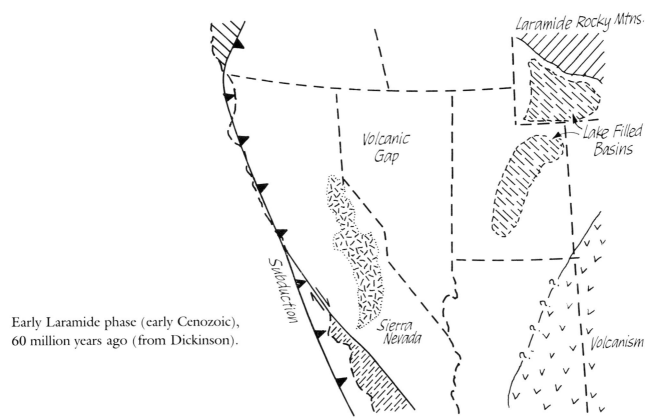

Early Laramide phase (early Cenozoic), 60 million years ago (from Dickinson).

velocity and the angle of underthrusting was the lowest, the effects of subduction would be felt as far east as the rising Rocky Mountains, more than six hundred miles distant from the plate margins. The activity had forty million years to deform and uplift the Rockies before plate movements would again slow down and terminate the Cordilleran mountain building.

The Colorado Plateau, bounding the Great Basin to the east, is an elliptical chunk of crust that acts as a discrete block of the continent. Rocks in the Colorado Plateau are almost undeformed by the Cordilleran orogeny. They acted as a solid, unyielding unit for the westward compressional forces of mountain building. The plateau rotated slightly clockwise and moved northward. The movement sheared the eastern boundary of the plateau in central New Mexico, central Colorado, northern Utah, and eastern Wyoming. The movement exerted great compression along this boundary. The Rocky Mountains, from Montana to New Mexico, resulted.

Laramide uplifts are clearly different from the earlier Sevier features. The Sevier is characterized by thin-skinned folding and low-angle thrust faults in what had been the Paleozoic continental shelf and slope of the eastern Great Basin. The Laramide Rocky Mountain uplifts, to the east of the Colorado Plateau, are large asymmetrical uplifts cored with basement rocks and bounded by higher angle thrust faults. They were formed by Late Cretaceous to late Eocene compressive forces on the thin cratonic shelf.

A Wave of Magmatism

As deformation moved easterly, so did arc magmatism. A wave of volcanism swept across the Great Basin from west to east. Eventually, as the subducting plate approached horizontal, the magmatism in the Great Basin waned in intensity and stopped. The mid-Paleocene to mid-Eocene peak of deformation in the Wyoming and Colorado Rockies coincided roughly with a prominent gap in arc magmatism within the Great Basin to the west.

Arrow Canyon, southern Nevada. Marine limestones deposited in the Paleozoic seas are moved eastward during the Sevier thrust-faulting phase in the Mesozoic. *John Running*.

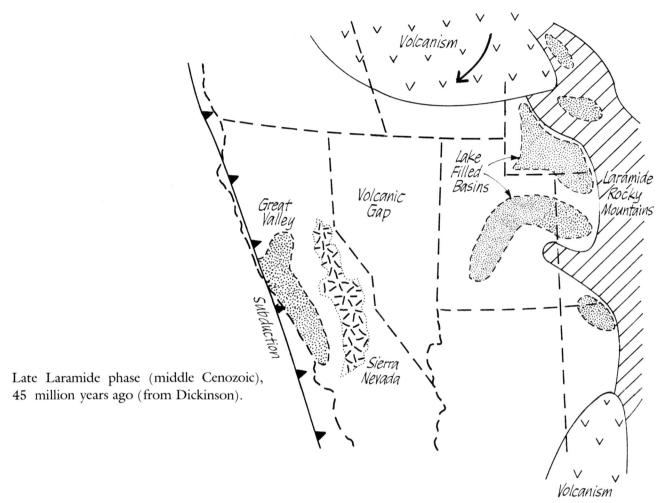

Late Laramide phase (middle Cenozoic), 45 million years ago (from Dickinson).

The End of the Cordilleran Orogeny

Simultaneous and monumental events changed the entire deformation pattern of the western Cordillera during the time period from fifty to forty million years ago. The Laramide phase terminated almost simultaneously throughout the western continent from Alaska to the Caribbean. Andean-style ocean floor–continent subduction ceased. The wave of arc volcanism abruptly terminated. These synchronous fundamental changes in the western Cordillera indicate a change in large-scale plate motions.

Far to the east in the Atlantic Ocean, geomagnetic stripes on the ocean floor may again contain the answer. The spreading rate of the opening of the Atlantic abruptly slowed down. During the same time period, forty million years ago, a large kink formed in the Hawaiian Island chain as the motion of the Pacific Plate changed to the northwest. As a result, the convergence rate of the oceanic Farallon Plate to the west and the North American Plate decreased by almost one-half. After forty million years ago, the rate has been less than three inches per year. This abrupt slowing down signaled the end of the Laramide.

The Laramide phase is coincident with, and probably caused by, the eighty- to forty-million-year time of rapid convergence between plates. Profound global plate tectonic changes bracket the Laramide phase.

Carson Sink, northern Nevada. The high ground surrounding the sink is marked by the horizontal shorelines of receding Ice Age lakes. *Tom Brownold*.

12 The Cenozoic

TWENTY-FIVE MILLION YEARS OF EROSION

After uplift comes erosion. Just as we begin to age from the earliest moments of birth, so the Laramide Highlands began to erode. Erosive forces were scouring the rocks from the earliest time of uplift. They soon revealed signs of the abrasiveness of age; the rocks were deeply beveled.

Erosion is the opposite of deposition. When rocks are being destroyed, no deposition is occurring at the site. Miles away, the accumulated detritus of erosion may pile up into new deposits, but at the site of destruction there is no record. The thief leaves no record except the empty shelf. Consequently, our knowledge of the early Cenozoic in the Great Basin is severely limited. The pages of the rock record have been torn out, and history is sadly incomplete.

The picture is still not complete and portions may have been forever erased. There are some scraps of information, like a few igneous rocks and a few out-crops of a clastic sedimentary unit in the central Great Basin. This sandy layer seems to have been deposited in a stream bed. Associated with it are some lake sediments. Perhaps this stream and the lakes were the result of internal drainage. There may have been some post-Laramide rejuvenation along faults—lithologic gerovital. The upthrust blocks may have obstructed streams draining the Laramide Highlands, creating a smaller version of the Great Basin long before the actual Great Basin was formed.

There were large volcanos to the east of the Great Basin during this early Cenozoic time. Our area, however, was so devoid of volcanism that geologists refer to the Laramide igneous gap when they refer to the first twenty-five million years of the Cenozoic in the Great Basin. This gap was soon to be violently and abruptly filled.

THE ENIGMATIC IGNIMBRITES

One of the most puzzling time periods in Cordilleran history is the middle Cenozoic. This was the Oligocene–Early Miocene time of about forty to twenty million years ago. The slow change of history was again punctuated by paroxysm.

Violent volcanic eruptions burst out in the northeastern part of the Great Basin a little more than forty million years ago. A great wave of volcanism engulfed most of the Great Basin, spreading from the northeast to the southwest. Huge eruptions of white-hot ash, lava flows, and ash-flow tuffs buried the landscape. The chemistry of the parent magmas is similar to those that had earlier created the granites of the Sierras. Here in the Great Basin, however, the molten material blasted out of volcanic vents. Fiery-hot clouds of ash and gases seared the land surface. Instead of becoming the intrusive interior of another Sierra Nevada, the magmas belched out onto the surface. The Great Basin was blanketed by the guts of what would have been a lofty mountain range.

The chemical nature of these eruptives created ash-flow tuffs comprised primarily of rhyolite, the extrusive counterpart of granite. Incandescent, white-hot ash falls from fiery volcanic clouds are termed ignimbrites. These rocks fused by their great heat into a welded tuff. The Defense Department tests some of its nuclear weapons on the Nevada Test Site in these hard, resistant rocks.

They are the dominant rock type in most of the ranges of the central and southern Great Basin. The volume of ash expelled is incredible—an individual ash fall might cover an area of four to five thousand square miles. One ash layer might be six hundred feet thick. In some cases, an original volume of just one ash-fall eruption might be from one hundred to

Clover Creek, Nevada. Ignimbrites stand as steep cliffs reflecting their dense inner structure. Volcanics comprise many of the ranges of the central and western Great Basin. *Tom Brownold.*

tered by these volcanic outpourings until only seventeen million years ago.

There were some lava flows associated with this ignimbrite time, and they are most abundant in a broad sweeping arc from central Nevada to central Utah. The lavas are generally andesites and dacites. We usually visualize lava flows as black basaltic rock, but these lavas were different, being typically brown, tan, or gray. Black flows were certainly atypical.

As in all catastrophic events, there are a few survivors that recall some of the events of the time. After diligent searching, geologists have discovered a few small depressions in the volcanic plain. These hollows formed between the volcanic eruptions. Fragments of stream-eroded volcanics washed into the swales and are interbedded as clastic sedimentary rocks with the silica-rich ash-flow tuffs. Some fresh-water limestone from ephemeral ponds has been located in Lincoln County, Nevada.

The enigmatic ignimbrite episode is a fascinating event rarely recorded in the geologic past. Such a profound change in geologic environments was most likely caused by geologic events initiated far beyond the Great Basin and was probably linked to a worldwide change in plate tectonics. But what might have been the change in plate movements to cause such a radical break from the past?

THE OCEAN SLOWS DOWN

Eighty million years ago North America separated from Eurasia. The action was accompanied by an acceleration of the North American Plate. Our plate began moving twice as fast. The change in pace from a leisurely three inches per year to the dizzying speed of six inches per year may not seem earth-shaking, but it literally was. We overrode the sea floor plate to the west so quickly that it was unable to sink down into the underlying slush and melt. The sea floor plate was close to its origin, the spreading center, and was young and hot. As it slid under North America at a low angle, the younger and hotter ocean plate came in contact with the underside of the continent. The heat and pressure of the underlying plate is thought to have created the easterly wave of volcanism and compression associated with the Laramide phase. The Rockies resulted from an acceleration of North America away from Europe.

If the easterly moving compressional and volcanic pulse of the Laramide was initiated by an acceleration of plate motion, perhaps it was terminated by an abrupt deceleration. Again, we go far from the Great Basin to search for the answer. Submerged deep below the cold waters of the North Atlantic, geophysical probes reveal magnetic stripes in the volcanic rocks of the sea floor. During a time interval from forty to about nine million years ago the opening of the Atlantic virtually ceased. The North American Plate stopped its westward drift.

As the impact velocity between the sea floor Farallon Plate and the North American Plate significantly decreased, the subducting slab would have had more time to cool before sliding under North America. Oceanic plates are formed by the solidifying of hot rising magmas at spreading centers. Consequently, the plate is hottest adjacent to the rifting center and cools and sinks with increasing distance from the spreading center. When North America stopped overriding the sea floor, the oceanic plate adjacent to the continent would have increasingly cooled with time. This cold plate must have contracted and increased in density. It would then plunge beneath the continent at a higher angle than the formerly fast-moving hot and expanded plate. Like a trap door slowly swinging down on its hinges, the contracting plate would steadily sink with a steeper and steeper angle as its density increased. The increased angle of descent would plunge the subducting slab deeper into the hot subcrust below the surface plates. Rapid plate melting intensified magmatism at the surface above the melting plate immediately above the region where the slab penetrated the subcrustal asthenosphere. As the trap door swung open, this locus of volcanism would shift westerly. Perhaps this is the cause of the reversal of the volcanic wave. The volcanic spasm that so greatly altered the face of today's Great Basin perhaps resulted from events in the North Atlantic.

Today if you hike into a Great Basin mountain range the odds heavily favor that you will be walking on Cenozoic volcanics—almost one-quarter of the surface of this region is ash-fall tuff or lavas. Much of the remainder is Cenozoic sediments eroded from the volcanics that now fill the valleys.

METAMORPHIC CORE COMPLEXES

Scattered throughout the western Cordillera are about twenty-five domal uplifts. These domes follow a sinuous belt from southern Canada to northwestern Mexico. Geologists have called these strange mountains of arched rocks metamorphic core complexes. They have stirred up considerable controversy among geologists of the western United States. Several conferences and symposia, a special memoir of the Geological Society of America, many technical articles, and countless late-night discussions have resulted since their first description more than twenty years ago. More than one-half of them have been discovered in the past ten years.

These domed uplifts are isolated from one another, but they generally have common characteristics. They have a core of highly deformed metamorphic and plutonic rocks overlain by an unmetamorphosed cover that has been stretched and detached from the underlying rocks. Most of the complexes either formed or reactivated during the time of the ignimbrite eruptions of the middle Cenozoic, approximately fifty-five to fifteen million years ago.

They seem to be associated with local pulling apart after the Sevier-Laramide compressional thrusting. Perhaps this tensional rifting resulted from the steepening dip of the downward-moving slab of the ocean floor plate, but the debate about their origin continues. The core of these gneissic intrusive complexes was apparently uplifted and cooled rapidly. The unmetamorphosed sedimentary rock cover slid off the uplifted core into the surrounding basins. This sliding was driven by gravity along large denudation faults. The combination of upbowing metamorphic core complexes and large-scale calderas caused by the evacuation of volcanics must have created a most startling landscape.

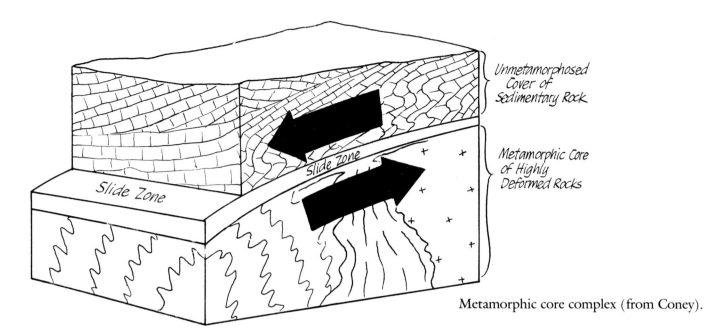

Metamorphic core complex (from Coney).

Three of these metamorphic core complexes are located within the Great Basin. Our oldest described Prepaleozoic rocks have been uplifted in one of the complexes. This is the Albion–Raft River–Grouse Creek Range complex, which lies in the common corner of Idaho, Nevada, and Utah. To the southwest by one hundred miles lie the Ruby Mountains, one of Nevada's most scenic ranges and another metamorphic core complex. Southeast by another one hundred miles along the Nevada-Utah border is the third complex. This is the Snake Range, crowned by Wheeler Peak, the second highest mountain in Nevada.

The Albion–Raft River–Grouse Creek complex has a core of ancient Prepaleozoic basement metamorphics overlain by possible Prepaleozoic and Paleozoic sediments. The sedimentary rocks are highly metamorphosed. The heating event is interpreted to have taken place between the Late Triassic and Middle Cretaceous times. Some metamorphism continued until the Miocene, indicating that uplift and shedding of the cover rocks may have continued until about eleven million years ago.

In the Ruby Mountains, the core is a highly meta-morphosed and folded complex of latest Prepaleozoic and Paleozoic rocks. These metamorphosed sedimentary rocks are thoroughly intruded by granitic dikes and sills. The granitic juices have so infiltrated the older metamorphics that about one-third of the core is igneous. Above the detachment zone of friction-heated rocks are the unmetamorphosed Paleozoic rocks. These were broken by low-angle faulting when the sliding brittle block was internally fractured during movement. A large granitic pluton on the south edge of the complex has been dated by rubidium-strontium methods as Middle Jurassic, and this may be the age of metamorphism and intrusion in the Ruby Mountain core. The faulting in the cover cuts igneous rocks dated as middle Miocene, indicating the time of latest movement.

The Snake Range has a core of low-grade metamorphosed late Prepaleozoic and Cambrian siliceous clastics and carbonates. These rocks are only lightly metamorphosed, although those that are adjacent to plutons have been baked more extensively. The age of the metamorphism is uncertain. It may be as old as Jurassic and perhaps continued until the Tertiary. The overlying cover is a nonmetamorphic pile of Paleozoic rocks. Faults that end downward in a low-angle slide plane cut the cover rocks. The slide zone, called a decollement by geologists, separates the core from the cover. The age of sliding seems to be in the Tertiary, about seventeen million years ago, but this date has some ambiguities.

THE RIDGE COLLIDES

Changes in plate geometry sometimes evolve quickly. The spreading center at the oceanic ridge system to the west of our continent was creating two plates: the Farallon to the east and the Pacific to the west. The Farallon was moving easterly and subducting beneath the North American continent. The Pacific Plate was moving northwesterly, opening up the Pacific Ocean and eventually subducting in the far western Pacific. Not only were the plates moving relative to each other, but so was the ridge relative to North America. Gradually it encroached upon the North American continent. Twenty million years ago the ridge collided with the continent.

The ridge progressively contacted the continent in a south to north direction up the continental margin. This zippered the ridge against the continent all the way north to the Pacific Northwest. The spreading center of the ridge was probably extinguished.

The northwesterly moving Pacific Plate was plastered up against the westerly moving continent. The differential motions between the Pacific and North American plates initiated a side-by-side moving fault system. The most prominent fault in this transform system is the San Andreas Fault. This transform system probably became active in late Oligocene when the oceanic ridge first contacted the continent. The present San Andreas Fault probably did not absorb most of the transform motion until the beginning of the Pliocene.

Remnants of the Farallon Plate continue to subduct under the North American Plate today. Off the coast of Washington, Oregon, and northern California, a remaining fragment of the Farallon Plate continues to slide under North America. The subduction of this remnant and its melting at depth created the Cascades. It also generated the violent andesitic eruption of Mount St. Helens in 1980. There will be continued volcanism in the Cascade volcanos until the zipper closes all the way. Another scrap of the Farallon subducts below southern Mexico and adjacent Central America. Arc magmatism continues in this region. Between northern California and Mexico arc volcanism no longer exists. Highly explosive andesitic volcanism is unlikely now in most of the Great Basin. Instead, basaltic volcanism related to tension is widely scattered throughout the region.

When the ridge collided with the continent in an act of lithologic suicide, the spreading center source of the Farallon Plate was extinguished. The remnant piece continued to move and sink under the Great Basin, but the great period of compression caused by massive subduction was over. Gone were the days when the Farallon would import exotic terranes from elsewhere to be plastered up against North America and add new territory to the continent. Gone were the compressive forces to uplift the Sevier or the Rockies. Gone was the plate digestion under the continent, belching up light-weight sea floor residues to create the Sierras or the ignimbrite flareups of the Great Basin.

This was the dawning of a new geologic age, with a new style of tectonism. Now the shearing motion of the Pacific Plate is slicing off a portion of the western part of North America. Los Angeles has begun its slow, lurching motion to the north, ultimately to be plastered against southern Alaska as a new exotic terrane in the Far North. Far to the south, the Caribbean Plate has formed as Central America shifts eastward, and subduction creates the Lesser Antilles. Volcanic island arcs are slowly rising, smoking and shuddering, above the blue Caribbean waters. Here in the Great Basin, a new fabric will be designed for the earth's surface. A unique landscape will form. This is the age of Basin and Range.

BASIN AND RANGE

The big change began about seventeen million years ago. The Basin and Range topography that characterizes virtually all the present-day Great Basin be-

Elko County. Ash from the eruption of Mount Mazama outcrops in the walls of a wash in eastern Nevada. The eruptions of Mount Mazama resulted in the caldera collapse seen at Crater Lake, Oregon. *J. F. Smith, USGS.*

Another hypothesis suggests that a narrow rising plume of deep mantle material rose upward due to its high heat and hence low density. Such plumes have been suggested to drive plate movement and to cause the heat rising beneath Yellowstone National Park. This idea has few adherents, since the evidence to support such a plume beneath the Great Basin is subject to much speculation and controversy.

A popular hypothesis relates basin-and-range structure to back-arc spreading. In this concept, the downward-moving subducting slab generates high heat from the friction between the plates, causing an upwelling of the heated lower density material. The rising magmas upbow and extend the overlying crust inland from the subduction margin. This model is particularly useful in explaining the basin-and-range characteristics of high heat flow, thin crust, and regional uplift and extension. To the north along the northern boundary of the Great Basin, great floods of basalt were erupted at essentially the same time as mid-Miocene extension in the Great Basin. These lavas may indicate deep ruptures in the extensional basin caused by back-arc spreading in the Pacific Northwest.

A recently proposed idea relates the extinction of subduction along the western North American boundary with basin-and-range extension in the Great Basin. As the North American continent overrode the sea floor plate to the west, subduction of the heavier sea floor occurred. Gradually North America approached the spreading center, now represented to the south by the East Pacific Rise. When the spreading center and the continent collided, spreading ceased and a transform boundary was established. Cut off from its source, the original subducting sea floor detached from the former spreading center. It continued moving down into the asthenosphere to the east of the new transform boundary. A slab window formed where the subducting slab detached. Here the hot magmas welled upward against the cold and rigid continental crust. The motion could have been the cause of the regional uplift and expansion in the Great Basin and

Models to explain Basin and Range (from Stewart).

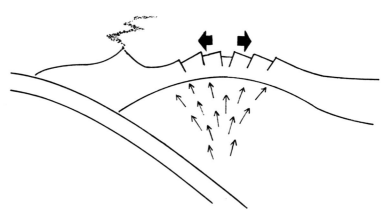

Back-arc spreading.

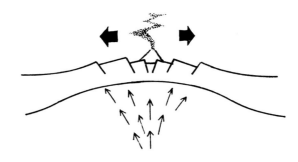

Mantle plume.

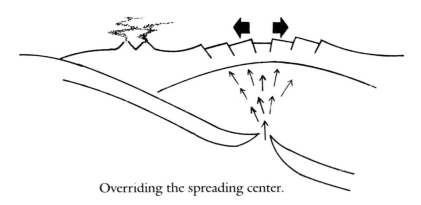

Overriding the spreading center.

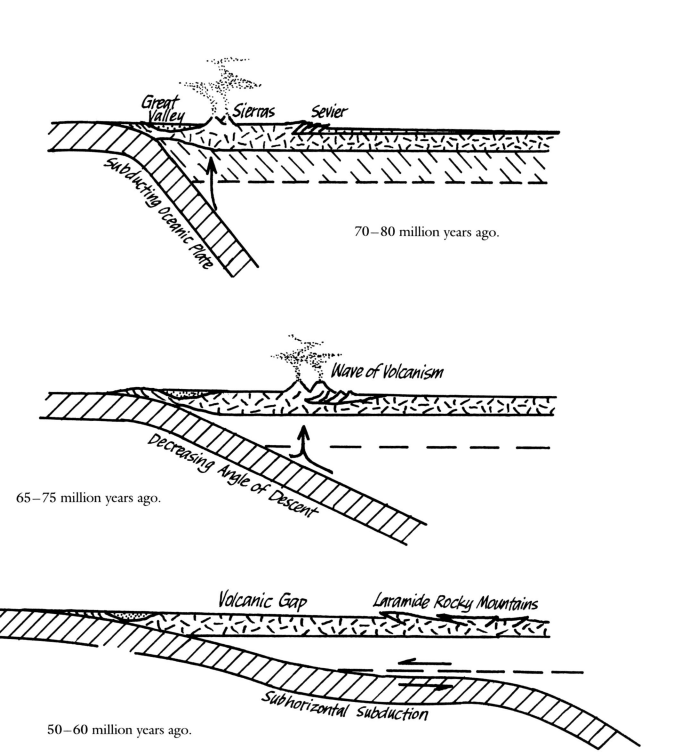

70–80 million years ago.

65–75 million years ago.

50–60 million years ago.

Latest Mesozoic—earliest Cenozoic (from Dickinson).

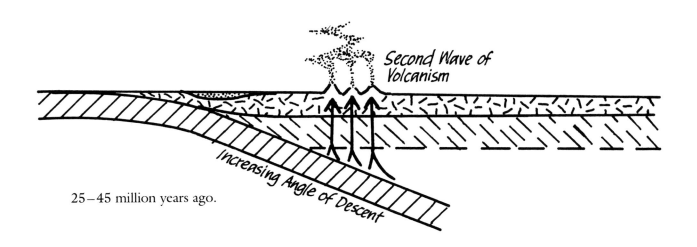

Second Wave of Volcanism

Increasing Angle of Descent

25–45 million years ago.

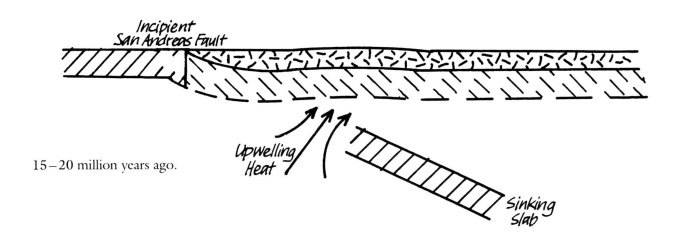

Incipient San Andreas Fault

15–20 million years ago.

Upwelling Heat

Sinking Slab

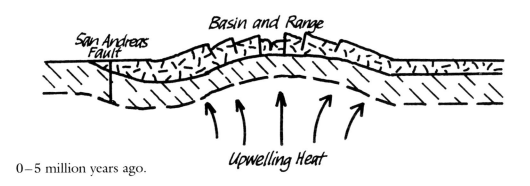

San Andreas Fault

Basin and Range

0–5 million years ago.

Upwelling Heat

Middle to late Cenozoic (from Dickinson).

could have created the basin-and-range structure. The slab window hypothesis is also difficult to apply to Mexico, where the continental crust has not yet contacted the spreading center but where basin-and-range structures exist.

Which of these ideas is most likely correct? Current geophysical and geological evidence certainly puts the weight onto those hypotheses that explain basin-and-range structure through plate movements. The largest support by geologists today seems to favor the back-arc spreading or slab window concepts. The answer is, we don't know. We must be aware of Thoreau's admonition against the tyranny of the majority. Perhaps the answer is yet to be determined. Every new idea at its inception will be alone.

VOLCANICS

Coincident with the onset of the formation of the Basin and Range seventeen million years ago, there was another profound change in the Great Basin. This one involves volcanic activity. The change is dramatic genetically, chemically, and visually. Iron-rich lavas, primarily basalt, poured over the landscape. These dark or black lavas extend from Canada to Mexico. They are best seen where they flooded the northern Great Basin. The rugged black mountains of southern Idaho, southeastern Oregon, and northern Nevada testify to the areal extent and dramatic beauty of these black lavas.

There was also a resurgence of andesitic volcanism. Andesites are widespread and voluminous throughout the western Great Basin. They are part of a continuous belt that parallels the western continental boundary from southern California and Nevada to northern Washington. These flows are often several thousand feet thick. The andesites are frequently capped by basaltic flows.

An east-west band of largely silicic ash-flow tuffs and small amounts of rhyolite extends across the southern Great Basin between latitude 37° and 38° north. These explosively emplaced rocks were erupted from a series of well-defined calderas in Nye and Lincoln counties, Nevada, and in adjacent California. Some of these calderas are so immense that they are best seen in photographs taken by spacecraft from five hundred miles above the earth. Southernmost Nevada at this time had thick andesite lavas and flow breccias with some smaller amounts of rhyolite. Large plutons invade some of these volcanics.

CATERPILLARS AND THE EGGBEATER

The dominant trend of the mountain ranges of the Basin and Range is slightly east of north. Many years ago a geologist compared them to dark fuzzy caterpillars crawling north. Seen from space, these dark tree-clad mountains indeed match the description. But peering down from our space platform, there is an obvious delineation in western Nevada where the caterpillars stop. This belt terminates the orderly ranges to the northeast from a disorganized geologic mess of discontinuous and arcuate ranges to the southwest. The confused ranges look as though a fifty-mile-wide eggbeater had moved through plastic rock. This zone is the Walker Belt, or Walker Lane.

The Walker Belt trends northwesterly parallel to the California-Nevada border. It extends for a distance of four hundred miles from the southern part of Nevada to easternmost California. Some geologists believe these ranges are contorted by very large side-by-side moving faults with a displacement of as much as 80 to 120 miles. The west side has moved northwesterly relative to the east side. This lateral displacement has also folded the ranges into great fishhook bends adjacent to the Walker Belt. Not all geologists agree with the idea of lateral fault displacement, but some profound lateral forces have surely disturbed these rocks and forced an abrupt change of geology and topography.

Some of the deformation in the Walker Belt may have begun as early as the latest Early Jurassic, and the zone has been shifting ever since. There is good evidence of Cenozoic faulting in strata adjacent to Las Vegas Valley that may be coincident with the Walker Belt. Here rocks fifteen million years old have moved as much as the underlying Paleozoic

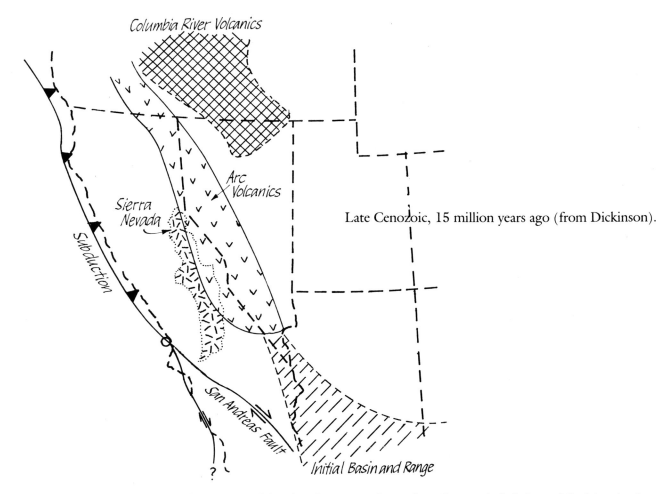

Late Cenozoic, 15 million years ago (from Dickinson).

rocks, but adjacent eleven-million-year-old volcanic rocks are relatively undeformed. Considerable motion on the southern portion of the shear zone must have occurred in this time interval between fifteen and eleven million years ago on what is called the Las Vegas Valley Shear Zone. Some geologists believe the belt is an ancient basement flaw that has been reactivated throughout time. The most recent large-scale lateral movements would have occurred in Late Tertiary time.

Total offset along the Walker Belt is difficult to determine. There is good evidence that ash-flow tuffs in Mineral County, Nevada, have been offset at least twenty miles. The Las Vegas Valley Shear Zone may have had twenty-five to forty miles of displacement. Faulting may not be continuous along its entire lateral extent, but the trend does seem to extend almost the length of the Great Basin.

Some late Cenozoic left-lateral faulting is also apparent in the southern part of the Great Basin in Lincoln and Clark counties, Nevada. A large late Cenozoic volcano has been neatly bisected by left-lateral faulting and shifted some twelve miles along the north shore of Lake Mead.

NORTHEASTERN WARPING

Northeastern Nevada and northwestern Utah also have some warping of Paleozoic sedimentary belts and two Paleozoic thrust belts. This bending occurs along a northwest-southeast lineament from just south of the Oregon-Idaho-Nevada junction through Wells, Nevada, and Wendover, Utah. The discontinuity has been described as the Wells Fault. The southerly block has apparently moved westerly relative to the northern block during Mesozoic time, thus disrupting the Paleozoic trends. The fault,

New Pass Mountains. Andesite volcanics cover a large portion of the western Great Basin. *Bill Fiero*.

Black Rock. Exotic terrane is represented by the dark rocks in this mountain in northern Nevada. These rocks sutured on to North America after a plate collision during the latest Paleozoic and earliest Mesozoic. *John Running*.

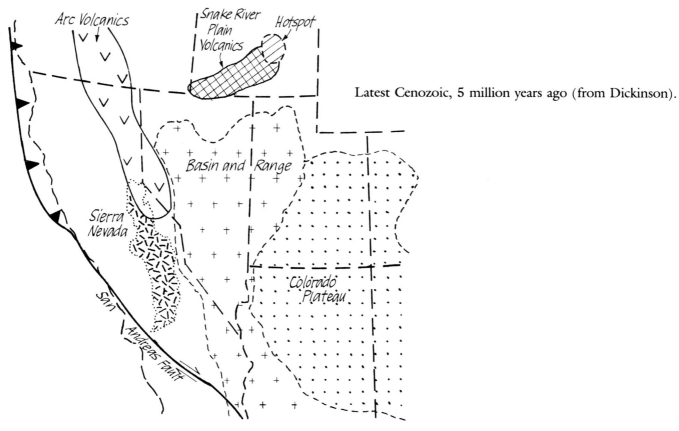

Latest Cenozoic, 5 million years ago (from Dickinson).

however, is not exposed and has been seriously questioned.

The latest interpretation considers the bending of the geologic units to be the result of a major flaw in the crust, perhaps even a bight, or slight curve, in the continental margin, along which vertical movement occurred several times during Paleozoic time. This hypothesis rests on the observation that there are large differences in the amount of offset of rocks of different ages that would not be reasonable if the movement took place after deposition of the rocks.

RECENT VOLCANIC ACTIVITY

There are a few places in the Great Basin where there has been volcanic activity in the geologically recent past. Basaltic cinder cones and lava flows with local sources occur along a trend from southern California through the western side of the Nevada Test Site to the Lunar Crater area of northern Nye County, Nevada. Here there are flows and cones

so recent that they appear to have been little altered by erosion. A deep pit, blasted out by volcanic gases, resembles a crater on the moon and gives the area its name.

The Mono Craters area at the foot of the Sierras along the western margin has had recent activity with obsidian flows, explosive eruptions, and caldera collapse. Recent seismic activity and new steam vents could presage more volcanism in the Mammoth Lakes area. Certainly this region could become active again.

Volcanism in the Lava Beds area of northern California and along the Snake River at the northern edge of the Great Basin is recent. On the southeastern margin of the Great Basin, along the border with the Colorado Plateau, there are Recent basaltic ash cones.

SIERRAN UPLIFT

The granitic core of the Sierra was emplaced about

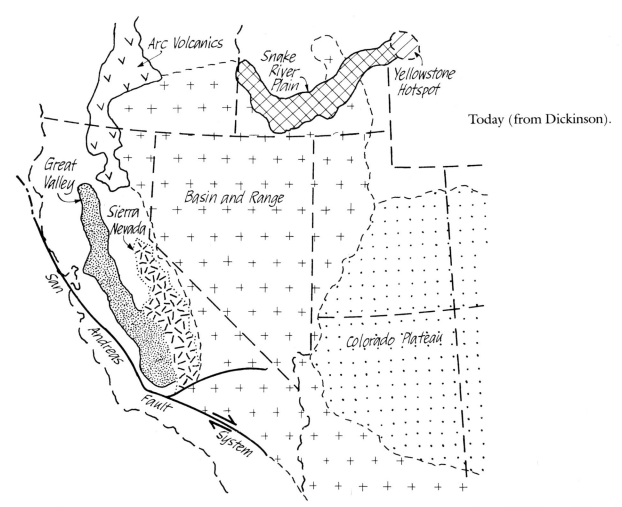

Today (from Dickinson).

210 to 80 million years ago. The Sierras were first uplifted shortly after the plutons intruded during the Nevadan phase of the Cordilleran orogeny. At that time, the range was perhaps three thousand feet high. Erosion ripped at the granitic uplift from about eighty to fifty-five million years ago. Cubic miles of rock were stripped from the layers that overlay the granite and from the granite itself. A thick prism of sediments, composed of the detritus from this erosional episode, was deposited along the California coast. The mountains were reduced to low hills.

About thirty million years ago in the middle Oligocene, violent volcanic eruptions buried the subdued range in ash falls and volcanic mud flows. In the northern part of the range, only a few isolated peaks projected above the volcanic mudflows after the eruptive holocaust. We are still in a period of volcanic activity, as attested by the earthquakes indicating movement of magma at Mammoth Lakes.

Earthquakes shuddered the Sierra, uplifting and tilting the block westward, beginning about twenty million years ago. Radiometric dating of the volcanics indicates an average of about one thousand feet of uplift per one million years for the past ten million years. About three million years ago, glaciation began to carve the mountain crests into the rugged scenery we associate with the Sierra today. The eastern side of the Sierras has been subjected to continued intensive faulting, uplift, and westward tilting during the past 2.5 million years. Uplift continues.

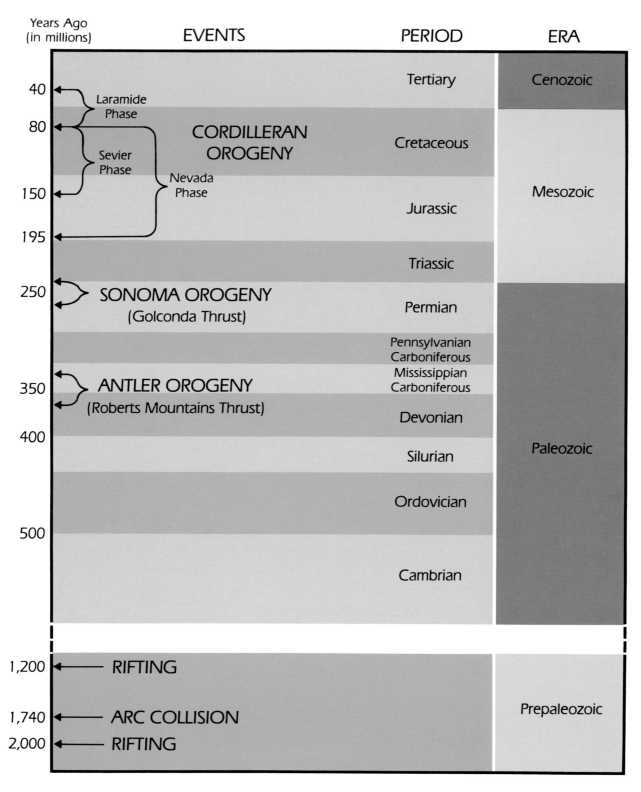

Time Chart of Earth Movements in the Great Basin.

PART III
The Great Basin Today

Lehman Caves. Uplift, coupled with the end of a period of high precipitation at the end of the Ice Age, has resulted in a lowering of water tables throughout much of the Great Basin. Solution zones in soluble rocks were left dry, allowing dripping water to form stalactites from cave roofs. *John Running.*

Blue Diamond. This major gypsum mine is located a few miles west of Las Vegas. *Union Pacific Railroad.*

13 Mineral Resources

WHEREVER YOU TRAVEL in the Great Basin there is evidence of former or present mining activity. In some areas the land is so pockmarked with holes and scrapings that it has the appearance of having been the home of gigantic gophers. Some of the largest man-made pits on earth have been dug out of the Great Basin rocks in the search for minerals. Much of the human history of the region is traced through the discovery of a glint of silver on this ledge or a streak of gold through that vein.

ORIGIN OF MINERAL DEPOSITS

Many mineral deposits are derived from magmas. Minerals have different temperatures at which they crystallize out of solution. As the magma cools, those minerals that crystallize at the highest temperatures form first. Thus metallic oxides, chlorides, sulfides, and native metals often separate from the rock-forming minerals in the early phases of crystallization. Sometimes these early magmatic ore crystals form while the rock minerals are still liquid and they settle to the bottom of the magma chamber as a segregated ore deposit.

Other ore minerals crystallize at temperatures well below those necessary for the rock minerals to form. These late magmatic ore minerals remain molten after the rock minerals have crystallized and turned into rock. The late crystallizing ores may solidify either at the site of cooling or they may be injected into the surrounding rocks.

The remaining fluids in the magma are concentrated as crystals form. Such brine fluids may contain carbon dioxide (CO_2), hydrogen sulfide (H_2S), hydrochloric acid (HCl), or sulfur dioxide (SO_2). These fluids often infiltrate and hydrothermally alter the surrounding host intrusive rock. If later solutions rich in metal chlorides encounter the hydrogen sulfide in hydrothermally altered rocks, the metal sulfides may be deposited. Such deposits are sometimes rich enough to be ore deposits. Thus are created syngenetic ore deposits, or deposits formed at the same time as the enclosing rock.

Just before the complete crystallizing of the magmas, high vapor pressure fluids may be concentrated to such an extent that they shatter the area above the intrusive. This creates fractures into which the residual solutions may be drawn by the regions of lower pressure. These solutions, if heavily mineralized, may form epigenetic mineral deposits, or deposits formed later than the enclosing rock.

Hydrothermal solutions that escape from the intrusive may be changed by contact with the surrounding rocks and become alkaline solutions. Or, if the temperature remains high enough so that the HCl remains intact, the solution may remain mildly acidic. The mineral substances in these solutions may replace the rocks through which they travel, thus creating replacement deposits.

There are other possible methods for the concentration of elements into sufficiently localized deposits to be mined. Sometimes bacteria are the cause of the precipitation of metal sulfides. In other instances, submarine volcanic exhalations may concentrate ore. Such submarine deposits have been documented occurring today in the depths below the Red Sea.

Frequently these primary deposits are not sufficiently concentrated in the precious minerals to be minable. Sometimes nature helps to concentrate these otherwise uneconomic deposits. A secondary process, such as weathering, may free the ore minerals to be transported either in solution or mechanically. The erosion process sometimes sorts the minerals on the basis of their density and durability.

Borax Hot Springs, Alvord Desert, Oregon. Subterranean pressures force water, heated by deep passage in the earth, to the surface. Hot springs are common in the Great Basin today and in the past. Minerals brought to the surface in solution account for some of the ore in the region's mining areas. *Stephen Trimble*.

Death Valley. Salt crystals grow in hypersaline springs and pools in the interiors of many valleys in the Great Basin. Interior drainage traps the minerals which would otherwise travel in solution to the ocean. *John Running*.

Tonopah.
Donna Gripentog McKay

Such deposits, called stream placers, usually form in drainages below the weathered ore deposits. In other cases the weathering and solution remove the uneconomic minerals and concentrate the ore in the residual soil. Desert gold placers commonly form in this manner as residual deposits.

Another secondary process, oxidation, may alter the nature of surface ore deposits. Minerals exposed to oxygen at the land surface or above the water table are often dissolved and carried downward below the water table. Here the different chemical environment can cause these oxidized minerals to precipitate or replace existing minerals. This process is called supergene enrichment and the resulting deposits are often exceptionally rich.

The many mineral deposits in the Great Basin result from a variety of geologic processes. The rich and varied geologic history of the region has created metallic and nonmetallic deposits of both great abundance and variety.

AGE DISTRIBUTION OF ORE DEPOSITS

The age of hydrothermal ore deposits can be determined by using radioactive age dating of the isotopes in the minerals that crystallized during the ore deposition. Armed with such dates, geologists can help place the formation of an ore body into the calendar of geologic history. Most of the ore bodies of the Great Basin formed from the Late Cretaceous to the Late Tertiary time. A few, however, are much older.

During the Late Devonian time, the approaching Antler volcanic island arc lay off the western margin of the continent in what is now central and western

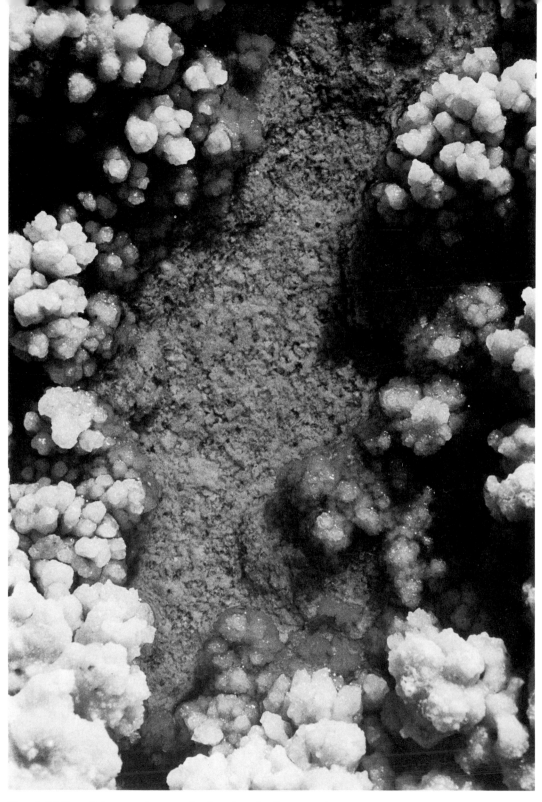

Death Valley. Salt crystals grow in the playa in the central part of the valley. *John Running.*

Nevada. Deep-water western assemblage sedimentary rocks lay between the island arc and the continent. These beds are associated with widespread bedded-barite deposits. Volcanism in the arc may have been a major source of the barium and sulfur necessary for the barite deposits. During the Late Devonian Antler orogeny, the bedded-barite deposits and their enclosing siliceous rocks were shoved eastward in great thrust sheets onto the continental rocks of central Nevada. Near Mountain City, Nevada, there are some massive copper sulfide ore bodies in Ordovician deep-water sediments. These may have formed at an earlier time on the sea floor in an environment similar to that of the Devonian bedded-barite deposits.

Mercury and antimony deposits are most abundant in the western part of the Great Basin. They generally occur in volcanic rocks or in deep-water siliceous sediments associated with volcanics. About 10 percent of the precious metals of the Great Basin are produced from these western siliceous deep-water rocks. At the same time, lead and zinc deposits were forming in eastern Nevada and western Utah in the shelf carbonate host rocks.

Eastward subduction of a sea floor plate beneath the North American continent occurred during Mesozoic time. The melting of the sea floor plate beneath the continent released light-weight constituents that intruded upward into the overlying continental crust. These intrusive magmas are thought to have been the source of the ore fluids that resulted in some of the largest ore bodies in the Great Basin.

Copper deposits in Nevada and Utah occur in close proximity to intrusive rocks of Jurassic, Cretaceous, and Tertiary ages. Iron deposits occur in the altered rocks adjacent to Jurassic intrusive bodies in east-central Nevada and near Jurassic-Cretaceous-Tertiary intrusives clustered in an elongate belt in western Nevada. Tungsten and gold deposits in Nevada are closely associated with Jurassic-Cretaceous-Tertiary plutonic rocks and are most numerous in western Nevada where these intrusive rocks are more abundant.

During much of Cenozoic time, silicic to andesitic volcanism occurred over large parts of the Great Basin as subduction continued on the Pacific margin of the continent. Andesites are the predominant host rock for precious metals in the Great Basin, with over 80 percent of the total gold production originating from these deposits. Some rhyolites of this same age are host rocks for several Great Basin gold-silver deposits. Subduction ended in the middle to late Cenozoic as relative motion between the sea floor plates and the North American Plate was taken up by transform faulting along the San Andreas system.

Crustal extension during and after the end of subduction created the Basin and Range. Volcanism associated with middle to late Cenozoic extension was fundamentally basaltic. These volcanic rocks have less than 10 percent of the Great Basin precious metal deposits associated with them.

ALIGNMENTS OF ORE DEPOSITS

Some geologists have recognized alignments of ore deposits in the Great Basin. These mineralized belts generally lie along trends that have been interpreted as major structural zones of earlier fault movement.

The Upper Cretaceous porphyry-copper deposits at Ely, Nevada, and gold-silver-lead deposits at Eureka, Nevada, lie within an east-southeast-trending regional belt of intrusives. The Oligocene ore deposits near Battle Mountain, Bullion, and Cortez, Nevada, are associated with intrusive rocks aligned in a southeast-trending belt in north-central Nevada. A second Oligocene mineralized intrusive belt trends east-northeast through the Bingham–Park City and Deep Creek–Tintic mining areas in Utah. Lower Miocene ore deposits, including those at Goldfield, Tonopah, Round Mountain, Wonder, and Pyramid, Nevada, occur in a southeast-trending zone along the Walker Lane and they coincide with a belt of upper Oligocene–middle Miocene volcanic rocks. A nearly coincident belt of middle Miocene ore deposits, the Comstock Lode, Rawhide, Camp Douglas, Silver Dyke, Manhattan, and Divide districts of Nevada, follows the same trend.

Stibnite. These beautiful antimony sulfide crystals were discovered in the White Caps mine in central Nevada. *Wendell Wilson.*

A second belt of middle Miocene ore deposits, the Majuba Hill, Seven Troughs, Ten Mile, Adelaide, Midas, and Jarbidge districts of Nevada, is aligned in a northeasterly trend. A southeast-trending belt of upper Miocene ore deposits links the Monitor and Bodie districts in California with the Aurora, Silver Peak, Gilbert, and Bullfrog districts of Nevada. This zone coincides closely with a belt of middle-upper Miocene andesitic and rhyolitic lava flows and breccias in eastern California and western Nevada.

PRECIOUS METALS

Gold

The Nevada Gold Province includes all the Great Basin region. Within it are some of the largest gold and silver deposits on earth. The first deposits discovered were relatively shallow veins, usually of Tertiary age, with some incredibly rich in gold or silver.

Many modern Great Basin gold mines contain ore that shows no color. The gold prospector of a hundred years ago would be amazed at the gold mines found in Nevada in the past thirty years. These mines are economically extracting gold that is colloidal in size and that cannot be seen by the unaided eye or even with a high-powered microscope. How many old-timers must have trod over some of Nevada's largest gold mines without the slightest suspicion of the riches that lay beneath their feet? How many of us today have walked over the gold mines of tomorrow?

Submicroscopic gold has been found in windows through thrust sheets in north-central Nevada. The deep-water western facies rocks of the lower Paleozoic were thrust easterly over the transition and shelf carbonate rocks during the Antler orogeny. The carbonates, and in some cases the crushed rock in the fault zone and the rocks above the fault, have been extensively mineralized. Several mines are associated with such thrust faults, such as the Carlin, Bootstrap, Gold Acres, Cortez, and Getchell deposits.

The Carlin mine, in the Lynn district of Eureka County about twenty miles northeast of the town of Carlin, Nevada, is a large open pit operation with large gold reserves. A more recent discovery, the Jerritt Canyon deposit of northern Elko County, may eventually exceed Carlin in production. The first discovery of gold in the Lynn district was in 1907, but the disseminated gold ore of Carlin wasn't discovered until 1962. Some geologists believe the Carlin deposit is the result of hydrothermal mineralizing solutions moving along highly inclined fault zones. The deposition of very fine gold occurred in the capillaries of fine clastics where carbonate was removed along the bedding planes. There seems to be a direct association of the gold with organic matter. The deposit may be of shallow low-temperature origin that possibly resulted from hot springs. Recent studies indicate that the ore fluid could have been derived from rainwater percolating to depth and perhaps enriched by mineral-laden water of deeper origin.

The Getchell mine has many similarities to the Carlin mine. The richest ore at Getchell was a hydrothermal replacement of wall rock along a steep fault. It is also thought to be a possible hot spring deposit.

A major gold deposit of Central Nevada is also of the disseminated gold type. The Northumberland deposit occurs in Paleozoic carbonate rocks that underlie a major thrust fault. It is broken by several steep faults as well. The deposit is spatially associated with the Northumberland caldera, a major collapse feature about thirty-two million years in age. Some of the minerals associated with the ore deposits have been dated at eighty-four million years, indicating that the ore is older than the caldera collapse and not genetically related to it. The Northumberland mine is currently active as a large open pit mine. The ore is hauled down to leaching facilities on the east side of Smoky Valley.

Another large central Nevada deposit is located at Round Mountain. This large open pit operation has been active since 1976. Although the deposit is of very low grade, it is perhaps the largest reserve of gold of any of the modern gold operations in the

Evidence of mining activities is found throughout the Great Basin. Virtually every range has been scoured by prospectors. *John Running*.

Great Basin, with over nine million ounces of gold estimated to be still in the ground. The Round Mountain deposit is probably related directly to a caldera located to the immediate west of Round Mountain under Smoky Valley. The host rock is a rhyolitic ash flow tuff. Mineralization has been dated at about twenty-five million years.

The Gold Mines at Manhattan in central Nevada are also probably related to a caldera. The host rocks of this deposit are Paleozoic carbonates, although the mineralization is apparently associated with the sixteen-million-year-old caldera. Manhattan was placer mined years ago, but some underground mining continues today.

One of the most outstanding gold districts in the Great Basin was the Goldfield district of Esmeralda County, Nevada. This district produced over $100 million in gold, with most of the gold produced during the short time period from 1904 to 1918. Tilted veins in late Tertiary volcanics contained gold. These exceptionally rich bonanza deposits were mined near the surface where the gold was deposited by successive precipitation in open spaces with only minor replacement. It is thought that the gold-bearing magmatic solutions were alkaline. The surface waters were acid, due to contact with hydrogen sulfide. The mixing of these two chemically different waters caused the near-surface precipitation of the gold.

Other gold deposits of past and some continuing interest include Tuscarora, Aurora, Jarbidge, and the Golden Arrow in Nevada and Mercur, Utah.

Silver

The production of silver has often paralleled that of gold, since silver is often a by-product of gold mining. Silver is also associated with lead, copper, and zinc and in many operations it is the silver that makes the mining of the other metals profitable. Well-known silver deposits of the Great Basin region include the Comstock, Tonopah, and Hamilton districts in Nevada and the Tintic, Bingham, and Park City districts in Utah.

The Comstock Lode was the scene of early west-ern mining and since 1859 has produced over $700 million in silver and gold from a scattering of mines along a trend several miles long. The bonanza period was in the 1870s, when the high grade ore was mined. Production has steadily declined, although some cyanide leaching of old dumps has been undertaken recently. The Comstock Lode is associated with a fault zone with about 2,700 feet of movement that separates Tertiary and older volcanic rocks from Mesozoic rocks. The fault zone is hundreds of feet wide and branches both upward and downward. Miners followed it downward for over three thousand feet but most of the bonanza deposits were found in the first thousand feet where supergene enrichment had occurred. The lode was flooded with hot waters at about three thousand feet. The rock temperatures of the lower levels reached over 85° F, which suggests an underlying uncooled igneous rock mass. In 1865 the Nevada legislature gave Adolph Sutro the right to drive a drainage tunnel beneath the Comstock from a point about four miles to the east of Virginia City. It was completed in 1878 after enormous financial and operational difficulties. The tunnel was a financial disaster since it was completed about the time the bonanza deposits were nearing exhaustion. Virginia City is still a thriving community, visited by thousands of tourists who come to see the original mansions, bars, and stores.

Nevada's Tonopah district was the last of the great bonanza silver districts in the Great Basin. It was discovered shortly after the turn of the century. Local folks like to tell the story that the silver lode was discovered when a prospector, Jim Butler, picked up a rock to chuck at his recalcitrant mule and the stone turned out to be a piece of silver ore. Since 1900 more than $150 million has been produced. In the early days, the bonanza silver ran over $200 a ton. Between 1911 and 1930 the grade fell to about $15 per ton. Since 1930 the ore has averaged around $10 or less per ton. The ore is within a sequence of Cenozoic flows, tuffs, and breccias that is probably related to a caldera. The chief control of the mineralization was a major fault with a displace-

Goldfield, Nevada. This is the view east from Florence Hill over the Atlanta claims in 1909. *F. L. Ransome, USGS.*

ment from 450 feet to almost half a mile. Other faults served as ore channels replacement veins.

Utah's Tintic district has produced over $600 million since its discovery in 1869. About one-half the value has been silver and the remaining half copper, gold, lead, and zinc. The rocks of the district were warped into a broad downwarp and broken by large thrust faults during the Sevier phase of the Cordilleran orogeny. The rocks are an assemblage of over twelve thousand feet of Paleozoic shelf sediments overlain by volcanics and intruded by magmas. Coincident in time with the igneous activity was a period of intense faulting. These faults are thought to have been the conduits for great volumes of hydrothermal solutions. The ores are thought to have been deposited from these solutions. Some of the ore bodies are replacements of the folded, faulted, and crumpled limestones of the Paleozoic.

BASE METALS

Copper

Some of the world's largest copper deposits are found in massive bodies of intruded magmatic material. These ore bodies are referred to as porphyry coppers, and they are the source of most of America's copper production. The largest copper mine in the United States is the Utah Copper Mine at Bingham, Utah. This deposit has produced in excess of $15 billion.

The intrusives associated with the Great Basin copper deposits are derived from the melting of the subducting Mesozoic sea floor plate under the continental borderland. The intrusives have heavily cracked margins, probably due to shrinkage during cooling or from the high vapor pressure of late hydrothermal solutions derived from the magma. The shattered shell was permeable for these mineralizing solutions, and the porphyry copper ore is usually found in the outer cracked portions of the intrusives or in the overlying broken rocks that have been heavily altered by hot water solutions. Thus a great volume of rock became sufficiently mineralized with copper to become ore. Later deep weathering and

erosion often released copper from early-formed sulfide minerals and secondarily enriched the ore, giving rise to a zone of high grade supergene sulfide and oxide ores.

Porphyry copper deposits are characteristically thick blankets sometimes more than one mile wide. Within these large zones, smaller oxidized high grade ore is mined in the earliest phase of production. The great bulk of the ore, however, is usually low grade due to the nature of the original mineralization and is often mined on a large scale with open pit techniques.

The Utah Copper Mine is located southwest of Salt Lake City. The intrusive mass penetrates upper Paleozoic limestones that have been heavily mineralized. Over ten million tons of metallic copper have been mined at Bingham Canyon, as well as large quantities of zinc and silver-lead.

The Ruth, Nevada, copper mines are associated with intrusions aligned east-west along a major fault system. Several of these intrusions are metallized and four have been mined. The Ruth underground mine, the Kimberly Pit, and the Liberty Pit are the major mines in the district. The Liberty Pit provided about 70 percent of the production of the district. The Ruth and Liberty Pit produced about 225 million tons of ore.

Lead and Zinc

The massive copper deposits at Bingham, Utah, have also produced over $650 million in lead-zinc during the last 125 years. Most of the ore is found in folded upper Paleozoic limestones and quartzites around the porphyry copper deposit. Waves of hydrothermal solutions moved outward from the igneous intrusives altering the limestones. The metals were deposited in successive zones outward from the intrusive margins: first copper, then lead-zinc, and finally the silver zone. The lead-zinc is mostly found in replacement deposits and fissure veins.

The Park City, Utah, district had prolific lead-zinc lode fissures and bedded replacements in limestones. The mining in the district is mostly inactive today but the region has recently developed another bonanza deposit. It has become a major ski resort, surrounded by condominiums, golf courses, and recreation areas.

FERROALLOY METALS

Manganese

This mineral is the most important alloy in making carbon as well as high-manganese steel. The presence of the mineral creates an exceptionally hard metal. A ton of steel usually requires from thirteen to twenty pounds of manganese.

The Three Kidds mine, just east of Henderson, Nevada, produced over two million tons of ore containing 20 to 25 percent manganese. The ore body lies in faulted late Cenozoic sediments. The mine, active until 1961, has been abandoned since that time.

Molybdenum

Practically all the molybdenum mined is used as a hardening alloy in steel manufacturing and increases both the strength and ductility of the steel.

The Hall property, twenty-five miles north of Tonopah, Nevada, contains about 150 million tons of molybdenum-copper ore. The Bingham porphyry copper deposit in Utah has produced more than 120 thousand tons of molybdenum ore per day as a by-product of copper production. The Great Basin has the potential of becoming an important source of molybdenum in the future. Two deposits now under development, Mount Hope in Nevada and Pine Grove in Utah, contain immense reserves of molybdenum.

Tungsten

The hardest known cutting agents, after diamonds, are tungsten carbide and boron nitride. Steel containing tungsten is hardened and can also cut other steels. Tungsten is also used for filaments in light bulbs.

Mill City is Nevada's largest tungsten district with four separate mines: the Stank, Humboldt, Sutton, and Springer. These mines are developed from vein-like deposits in thin limestone beds of a western facies siliceous sedimentary rock cut by a late Mesozoic intrusive. The ore was emplaced by replacement. This process entails essentially simultaneous capillary solution and deposition by which a new mineral is substituted for one or more earlier formed minerals. A recent revival of activity has consolidated operations under the name of Springer mine, and large reserves of tungsten ore have been developed.

The very rich tungsten ore in the Oreana, Nevada, deposit is associated with thin vertical or lens-shaped pegmatites. These intrusives, composed of very coarse crystals, are smaller projections from post–middle Mesozoic large intrusive bodies.

The Silver Dyke deposit, near Mina, Nevada, is found in fissure veins that are filled with the tungsten ore scheelite, silicates, and sulfides. The veins are the result of hydrothermal solutions that emanated from lower Mesozoic volcanics and intrusives.

Recent activity at the old camp of Tempiute, Nevada, has developed new large reserves of tungsten ore from two small Laramide intrusives, each about one-half mile in diameter. As at Mill City, tungsten ores at Tempiute have formed in limestones by a replacement process.

Other deposits of tungsten are found at Nightingale, Paradise Range, and Golconda, Nevada. The Golconda tungsten deposit is a hot spring deposit and is associated with manganese and iron.

OTHER METALS AND RELATED NONMETALS

Antimony

Antimony has the unusual characteristic of expanding rather than contracting when it cools. It is thus used in alloys, particularly for lead. When cast the alloyed lead does not change size. Antimony is also used for electrical and military purposes.

Small production has occurred in the Great Basin, primarily related to vein deposits associated with gold mineralization or with tungsten and mercury.

Barium

The major use of this metal is in the petroleum in-

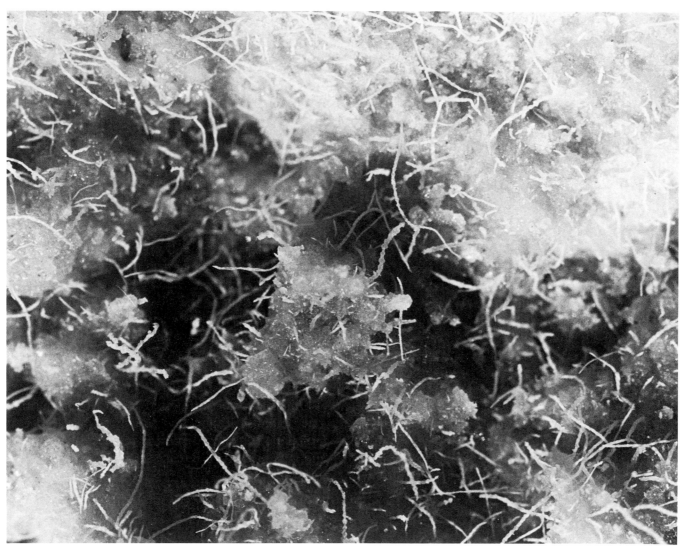

Death Valley. Fibrous salt crystals grow outward into cavities on the salt pan of the playa in the central part of the valley. *John Running*.

dustry to form drilling mud. This heavy fluid prevents blowouts by maintaining a higher pressure in the borehole than in the rock formations. A secondary use of barium is in the electronics industry.

Large sedimentary barite deposits were discovered by the U.S. Geological Survey in Nye and Lander counties, Nevada. These bedded-barites are located in early Paleozoic western assemblage siliceous rocks. The barite is thought to be the result of inorganic and organic chemical processes of deposition and is postulated to result from submarine volcanic exhalations. In 1980, Nevada supplied 85 percent of the United States barite production. Barite is mined by open pit methods.

Beryllium

Beryllium is a light metal that imparts high strength and fatigue resistance to copper, cobalt, nickel, and aluminum. Spor Mountain, Utah, is the largest source of beryllium in the western world. The beryllium in this deposit is disseminated in a Tertiary water-laid tuff that has been heavily altered by hydrothermal fluids. Large beryllium deposits are also known to occur near Mount Wheeler in eastern Nevada, but these are not presently economic.

Bismuth

This metal is used primarily for cosmetic and medicinal preparations. The smoothness of its salts makes them desirable for cosmetics. Some also help cure wounds and soothe digestive disorders.

The Tintic and Park City, Utah, mines have produced this metal in association with lead-zinc replacement deposits. These districts are among the most important sources of bismuth in the United States.

Cerium

Cerium, alloyed with 30 percent iron, forms ferrocerium. When this brittle alloy is scratched it emits sparks and so is used as "flints" or sparkers for cigarette lighters. It is also used in photography, glassware, and electronics. A main source is the Mountain Pass mine along the southwestern edge of the Great Basin in California.

Mountain Pass is the largest rare-earth metal mine in the United States. There are seventeen of these chemically similar metallic elements. They are not earthy and some are not rare. The elements usually occur together in groups of four to eight in one mineral. The principal mineral at Mountain Pass is bastnasite, which contains cerium, lanthanum, neodymium, and praseodymium, but also small amounts of samarium, gadolinium, and europium. Even though this area was intensively prospected during the past hundred years and many small silver, gold, lead, zinc, and copper deposits were found, the existence of rare earth metals was not recognized until 1949. All of the area containing economic rare earth deposits was acquired in 1950 and 1951. The average ton of ore at Mountain Pass contains 140 pounds of rare earth metals, of which half is cerium. It is mined by open pit methods.

Fluorspar

Fluorspar is a critical mineral in the production of steel since it fuses with impurities and removes them from the steel by forming a slag. It is also used to make hydrofluoric acid, necessary for etching glass.

There is small production in the Great Basin, but it deserves brief mention since it seems to have a genetic relationship with both gold and molybdenum deposits.

Lithium

Lithium is the lightest metal known. It is used in pharmaceuticals, batteries, photography, ceramic and glass industries, in many manufacturing processes, and has a major potential for use in the lithium battery. It is also increasingly being used as an alloy for light airplane metals.

The most important reserves in the United States are those associated with evaporite deposits. In the Great Basin it is found as a by-product from the brines of Searles and Great Salt Lake. The Great Salt Lake brines contain thirty-four to fifty-eight parts per million of lithium, although it is not yet utilized economically.

The largest source of lithium and about 50 percent of the United States reserve is found at Silver Peak, Nevada. Production is from brines below a large playa.

Magnesium

Magnesium is another very light metal. Its primary use is for the manufacture of light alloys that are used in airplanes and automobiles. It was also used for the manufacture of incendiary bombs in World War II. Magnesium is derived from natural brines, seawater, and rock-bound ore deposits. Consequently, it can be produced almost anywhere.

The Gabbs, Nevada, deposit has been the primary domestic producer of magnesite. The discovery was made in 1927, drilled in the 1930s, and was expanded as the result of the demand for magnesium metal for World War II. The world's largest magnesium plant was constructed during those years at Henderson, Nevada. This site was chosen since it is located only fifteen miles from the electrical power source at Hoover Dam. Much of the magnesite ore that was reduced to magnesium metal at this plant was derived from the Gabbs deposit. The ore body at Gabbs has proved reserves of over forty million tons of magnesite. Production to date has amounted to more than 95,000 tons of ore.

Mineralization at Gabbs occurs in a Triassic carbonate formation that was deposited in a shallow marine environment adjacent to the accreted Sonomia microplate. The light-colored, coarse-grained, massive dolomite that was deposited in the shallow water is intergrown with magnesite. It has been suggested that the magnesite is sedimentary and formed contemporaneously with the dolomite in this embayment of the Triassic sea. There is, however, an alternate hypothesis. The Gabbs magnesite bodies are located near a small granodiorite intrusive. There is evidence that the post-Triassic intrusion of the igneous body modified the magnesite bodies. Perhaps the magnesite deposits were formed by hydrothermal alteration by hot water solutions derived from the intrusion.

The waters of Great Salt Lake, Utah, have one of the highest magnesium contents in brine water in North America. The solar evaporation of these waters could constitute a major source of magnesium. The brines have a dissolved mineral content that would have a gross market value of about $75 million. They also contain significant and economically recoverable amounts of boron, bromine, and lithium. These brine deposit operations are still in the development stage but could become the primary source of domestic magnesium production.

Mercury

Mercury, or quicksilver, is a unique metal in that it is a liquid at room temperature. It is an essential mineral today because it is used in electrical apparatuses, pharmaceuticals, batteries, catalysts, and in agricultural, manufacturing, and scientific industries. The United States imports most of its mercury requirements today.

The Great Basin has been a producer of mercury. Most of these small deposits were discovered in western Nevada and are associated with either calderas or hot springs. These occurrences are often related genetically to some of Nevada's gold deposits.

NONMETALLIC MINERAL DEPOSITS

Most nonmetallics are familiar to all of us. This is the realm of such deposits as sand, gravel, and gypsum. The economics of such prosaic commodities are often misunderstood. The gross value of all nonmetallic products annually greatly exceeds that of metallic ores. Most nonmetallics are abundantly distributed throughout the world, so their value depends less on the material than on access to markets. The economic value of a nonmetallic deposit differs significantly from a metallic deposit in that it is determined largely by the cost of transportation. Metallic ores are usually refined to a pure form, whereas the nonmetallic minerals are used essentially in the form in which they are mined.

Bentonite

Bentonite is used in making heavy muds for the drilling of water and oil wells. It swells in volume

Death Valley. Delicate salt crystals and fibers grow after rains flood portions of the valley. Evaporite deposits constitute an important mineral resource in the Great Basin. *John Running*.

fifteen to twenty times when wet with water. Consequently it seals the walls of wells and prevents fluids or gases from entering the well bores during drilling. Bentonite consists mostly of montmorillonite clays and is a product of the weathering and alteration of volcanic ash or tuff. The alteration probably began while the fine-grained ash settled through fresh water. Bentonite clays are mined in the Ash Meadows area northwest of Las Vegas.

Diatomite

This sedimentary rock consists of siliceous shells of the microscopic plants called diatoms. It resembles chalk or clay but is mostly silica. It is also known as diatomaceous earth. It is so light that when dry it will float on water. It is used as a filter and filler. Diatomite is mined in several areas of the western Great Basin.

Glass Sands

There are scattered and locally important deposits of pure sand that are used in the manufacture of glass. Some of these sands, such as those in southern Nevada, are derived from Jurassic windblown deposits, while others are associated with sands concentrated during the recent water-rich environment of the Ice Age.

Gypsum

Some of the most lucrative mining operations in the Great Basin are associated with the extraction of gypsum for the production of wallboard. Some of the gypsum deposits are associated with the late Paleozoic retreat of the ocean from the continental margin and the deposition of bedded gypsum along the desiccating margins of the sea. Other gypsum concentrations are the apparent result of groundwater solution and redeposition of the earlier gypsum deposits during Mesozoic or Cenozoic times. Large open pit gypsum mines are located in the southern Great Basin adjacent to Las Vegas.

Lime

Lime is mined adjacent to the larger metropolitan areas in the Great Basin for the manufacture of cement. The usual source for the lime is pure beds of limestone deposited in the Paleozoic seas along the margin of the continent.

Sand and Gravel

These nonmetallic deposits are mined adjacent to every urban area in the Great Basin. The materials are usually derived from the erosion of rock or from volcanic ash. The deposits are often concentrated by the action of streams.

Semiprecious Gems

Precious gems are not known to exist in the Great Basin, but there is an abundance of semiprecious gems. These materials support a large population of rock hounds who prowl the rugged mountains of the region every weekend and create beautiful jewelry with their finds.

Turquoise is a copper aluminum phosphate that is created by descending cold waters. It is discovered in veins and nodules in a great variety of host rock. Nevada has historically provided most of the turquoise used in the Indian jewelry sold throughout the Southwest. The mining of turquoise is typically a small-scale operation, as the deposits are usually localized and scattered. One of the more significant deposits is the Wood or Crescent mine, Clark County, Nevada. This mine was worked by Indians, and artifacts at the site have been dated from the late 1200s. It was rediscovered in the late 1800s and is still a favorite weekend haunt for local gem collectors. The Royal Blue mine in Esmeralda County has been one of Nevada's most important mines.

Opal is composed of silicon dioxide and water. The fire opals show a play of colors, probably due to a refraction of light from thin layers within the opal. It is of secondary origin and is found in veinlets and nodules. The Virgin Valley opal field in Humboldt County, Nevada, has been an important producer from numerous shallow open pits. These opals are associated with petrified wood as casts of limbs and twigs and as veins filling cracks. The Roebling Opal, a 2,665-carat find, was discovered in this field in 1917.

Other semiprecious gems and stones are eagerly sought throughout the Great Basin. These include jasper, chalcedony, chrysocolla, hematite, rutile, and a myriad of other beautiful stones.

BRINE DEPOSITS

Playas and saline lakes are common throughout the Great Basin. These are the residual dry lakes or puddles from the blue water lakes that once filled many of the valleys in the Basin and Range. There are hundreds of square miles of level, glistening white salt beds to the west of Great Salt Lake. Almost every valley in the region has salt or sand in its interior portion and a few still contain water.

Alkali lakes, containing high concentrations of sodium carbonate and sodium sulfate, are common in the Great Basin. Owens and Mono lakes in California and the Soda Lakes in Nevada are examples of this kind of lake. The Soda Lakes contain about one-fifth each of sodium carbonate and sodium sulfate and three-fifths of sodium chloride. The source of the high amounts of sodium are the many igneous and some sedimentary rocks of the region that are rich in sodium. Potassium carbonate may be present, and common salt is always found. Calcium carbonate is usually precipitated out of these lake waters as limestone in the early stages of concentration.

Bitter lakes contain water with large concentrations of sodium sulfate but carbonate and chloride are also usually present. The sulfate is derived from the erosion of marine rocks that contain sulfates. Soda and Searles lakes in California are bitter lakes. The Soda Lake playa salts contain 88 percent anhydrous sodium sulfate.

Nitrate lakes are also found in the Great Basin. Soda niter, along with ammonium compounds, is known in Searles Lake and in lake deposits in Utah, Nevada, and Death Valley.

The most important lake source of potash in the United States is Great Salt Lake, Utah. Potash is also present in Mono and Searles lakes, California, and Columbus Marsh in central Nevada.

Borax

Borax is a well-known household commodity but it is of greater importance in industry. It is used as a cleanser, in pharmaceuticals, in the manufacture of paper and textiles, in metallurgy, in glass production, enamels, fire retardants, and gasoline additives.

Borax was first obtained from playas and borax marshes in the Great Basin. There are horror stories of the Chinese laborers working under the desert sun on the dry lake beds of Nevada digging the borates by hand. The mining of these impure products was later displaced by the extraction of bedded deposits of purer colemanite and ulexite from Death Valley. These beds are still the most important source of borax in the United States, but borax is also derived from brines underlying desert lakes. The brines are pumped from the interstices of the beds below the surface. Some of the hot springs and fumaroles of the Great Basin also contain borates.

The original source of the boron may have been spring waters, followed by concentration and evaporation. Many chemical and mineralogical transformations occurred before finally forming the borate minerals now found.

At Searles Lake, California, brines are pumped from below a playa and the various constituents are separated by evaporation followed by fractional crystallization. California bedded deposits are found near Kramer, Searles Lake, Boron, and in Death Valley.

ENERGY RESOURCES
Hot Water and Steam

The Great Basin has been stretched by Basin and Range extension. The crust, therefore, is thin beneath the region and it is a shorter distance down to hot subcrustal rocks than the average. Consequently, there is a high geothermal gradient beneath the Great Basin and this has made the region an interesting prospect for the geothermal production of energy. Rainwater percolating down beneath the basins encounters high heat at a relatively shallow depth and in some places is turned into steam. Sev-

Railroad Valley, Nevada. Oil production has not been significant in the region. The geology is favorable for the origin and entrapment of hydrocarbons, but exploration is difficult due to the complex earth movements. Gravel-filled valleys also mask the underlying geology. *Bill Fiero*.

eral hundred wells have been drilled to discover and delineate high temperature geothermal steam prospects. Lower temperature prospects have been explored for projects such as alcohol production, food drying, commercial and residential district heating systems, and for space heating in single-family residences. The future of such projects is promising.

Oil and Gas

Several hundred wells have been drilled in the Great Basin in the exploration for oil and gas. Only a few oil fields have had significant production. The most important of these are the Trap Springs and Eagle Springs fields in Railroad Valley, Nevada, and the new Blackburn field in Pine Valley, Nevada.

The region has excellent reservoir rocks to contain the petroleum, and some source beds in which the oil might have formed. There are ample folds and faults to entrap the hydrocarbons. The difficulty in petroleum exploration in the region has been the extreme complexity of the geology. There has been so much tectonic deformation that the potentially entrapping structures are broken by many faults. Basin and Range deformation has also downdropped huge basins that have filled with gravels from the surrounding uplifted blocks. The gravels mask the underlying geology and hide the potential petroleum-bearing structures. The uplifted ranges have well-exposed geology, but any oil or gas formerly present would have drained out and been lost to the atmosphere long ago.

The development of more sophisticated geophysical methods of exploration, particularly digital seismic techniques, has resulted in a rebirth of exploration interest in the Great Basin. There is more oil and gas to be found in the area but the search will be costly.

Coal

There are only a few commercial deposits of coal found within the Great Basin and these are primarily along the eastern margin. Cretaceous beds contain some coal seams between layers of sandstone and shale.

Uranium

Igneous rocks and uranium have a long and intimate association. The Great Basin certainly has no paucity of igneous rocks and consequently there are more than five hundred known occurrences of uranium in the region. More than half of these are related to the waves of volcanism that swept across the region in Cenozoic time. About half of the remainder are found in association with the Mesozoic intrusive igneous rocks.

The uranium in the igneous rocks is usually too disseminated to make a commercial deposit. Nature, however, has a technique for concentrating the ore that has been very effective. The volcanics were uplifted during Basin and Range faulting, beginning seventeen million years ago, and exposed to weathering. This erosion released the uranium from the glassy trap of the volcanics. Oxygen-rich groundwater, circulating through these weathered rocks, incorporated and carried the uranium. The high topography of the uplifted volcanics caused active groundwater flow, which gave an added impetus to the movement of the uranium.

When the waters reached an environment without oxygen, the uranium was precipitated and concentrated. Sediments rich in organic matter often became the hosts for such precipitation. During the past few million years the semi-arid conditions of the uplifted volcanics supported little vegetation. This was fortunate since plants might otherwise have entrapped the uranium ions in plant tissue. The ions moved with the water to the margins of lakes where there were abundant sediments rich with organic material. Upon reaching the reducing conditions of the rotting mass, the uranium precipitated out of the groundwater. Sometimes these concentrations are commercially significant today. There are other techniques by which uranium is concentrated but this is the most common within the Great Basin.

This method of precipitation of uranium is fundamentally different from that of most other precious resources. Consequently, uranium is usually not found in association with other metals. There are only about one-third to one-fourth as many uranium occurrences in the Great Basin as that of the base and precious metals, respectively.

Echo Wash, southern Nevada. Veins of gypsum outcrop in the walls of the dry wash. These selenite crystals reflect the sun like mirrors. *John Running*.

14 Water

THE GREAT BASIN is a land of contrast. The outline of the driest region in the United States is defined on the basis of the flow of surface water. The longest river of the region, the Humboldt, rises within the confines of the Great Basin, flows 330 miles to the west, and finally dries up before leaving the region. In fact, all the major rivers of the Great Basin—the Humboldt, Carson, and Truckee—begin and end there.

Many of the deep valleys are floored by monotonously flat mud-surfaced playas, or dry lake beds, that are usually shriveled and parched. Yet the flanks of the mountains bounding these desiccated valleys are etched by the shorelines of deep blue-water lakes that once filled most of these valleys. Such are the contrasts in water in this land of contrast.

Flash Flood

I remember a night camped under a deep blue-black sky perforated by brilliant sparkling stars. The black outline of a vertical sandstone cliff hung over me like the confining borders of a giant coffin. I chose a small alcove on the cliffside for my overnight home. It afforded a pleasant twilight view down the steep-walled canyon to the Colorado River tumbling southward through Cataract Canyon one mile distant. The night air was soft, with light, wandering, cool breezes ruffling the sleeping bag. I soon drifted off to sleep. Within a deep dream, a resonating rumble became part of a train ride over a high trestle. The sound slowly crescendoed and finally penetrated my dream with such authority that I lurched into wakefulness. The ground shook, and the parabolic walls of my alcove focused the cacophony onto my sleep-dulled brain. It was like a continuous roll of thunder or the insane pounding of snare drums by a demoniac drummer. The sound filled everything.

Through the pale starlight a surging mass plunged downcanyon. With a fluid solidarity it moved toward me. Occasionally the falling rounded contours of its face were pierced by branches or a tree that briefly leaped ahead of the surging mass and fell to the foot of the moving wall. The sound and sight were almost beyond my reason. Fortunately, my alcove was above the flood's crest, since I was transfixed in my sleeping-bag cocoon watching the tumult approach and fill the canyon below me.

The land of water contrast. The day had been warm and clear. Far to the east, black thunderclouds had climbed. They had grumbled and slashed lightning threats all afternoon. Hours later and miles distant from me, the storm-released waters had gathered together to create the desert flash flood.

By dawn, a muddy-slick surface coated everything below me and a cocoa-brown flow still filled the central channel of the canyon. In the land of drought, the respite becomes almost gluttonously extreme. I have viewed scores of these incredible events since and am always transfixed by each as though the sensory overload saturates my mental capacity for the unreal.

The lessons of flash floods are important to remember. Too many residents of the Great Basin are new to the region and only familiar with the urban environment. Sudden destruction can roar down a wash in either the desert or a city. Urban areas are faced with a difficult choice. Do you use flood control methods sufficient to handle the usual flood, or do you build oversized structures to handle the worst possible event?

Hydrologists discuss flood events in terms of their recurrence intervals. This is a convenient statistical measure of the threat of a flood. A ten-year flood is the largest flood you would expect every ten years. These are determined by studying the record of past

Mono Lake. Salt rims the shores of the lake. Until recently, the waters have been receding due to the diversion of tributary streams for human use in southern California. *John Running*.

flooding. A one hundred–year flood is likewise the largest flood you would expect to occur during a one hundred–year interval. Since it is a statistical measure, it does not mean that once you have had the one hundred–year flood you are safe for another ninety-nine years. The statistics always recycle, such that you have the same chance tomorrow as you had today. You can have two one hundred–year floods only a few days apart. Stated another way, there is a 10 percent chance in any year for a ten-year flood and a 1 percent chance for the one hundred–year event.

It is vital to understand, however, that a one hundred–year flood is not merely ten times larger in size than a ten-year flood. It is more likely a hundred, or even a thousand times larger. So what do you do to prevent urban damage? To build for a ten-year flood is certainly prudent. The one hundred–year flood, however, may require a channel a hundred or more times larger. The cost of such a channel for such an infrequent event becomes very expensive to raise by tax revenue or bond issue. We gamble with our future. Rarely is there a channel designed in our Great Basin urban areas for a flood with a frequency of one hundred years.

Great Basin Modifier

Water in the Great Basin is both abundant and scarce. Surface water, except in the montane regions or in the lakes, is a rarity that controls desert life. Edward Abbey described desert life as either being able to stink, stab, or sting. Since the bodies of most plants and animals are primarily water, life here usually protects its fluid resource beneath a surface armor.

Water is simple. H_2O. The simplicity of the chemistry of water is deceptive. The simple formula speaks of its nonspecialization. Specialists are restricted in what they can do and how they function. Not true for the generalist. The range and scope of activity and ability has few limitations. So it is with water.

Since water is simple, it has an almost universal attraction to other things chemical. Just as a simple, straightforward statement usually disarms complex reasoning, so does simple water have the uncanny ability to take apart more complex substances and dissolve them. Water is the closest substance we have to a universal solvent.

Water is a great solvent because it is a dipolar molecule, which means it can dissolve most other chemicals with ease. With a positive and negative pole to attract its opposites, many substances succumb to its siren call. This dipolarity reduces rock. Success lies in being attractive, simple, fluid, and patient. Virtually all rocks are eventually soluble in water, even in the Great Basin desert.

Some rocks yield more readily. Carbonates, originally taken from water in seas or lakes, succumb quickly. In the Great Basin, however, surface water is rare and limestones stand defiantly as vertical cliffs against the infrequent summer shower. Deep under those proud cliffs, though, the simple disorganizer is rotting the foundation. Groundwater dissolves passages and caverns and takes apart the strata grain by grain, weakening the entire mountain edifice. Water will win in the end, of course.

The crust is stretched thin by tectonic forces in the Great Basin, allowing water to percolate into heated subcrustal regions. The hot water, returning to the surface, may combine its dissolving forces with heat in order to redouble the erosive attack. Sometimes the addition of fluid excesses from rising magmas add heat. Ions derived from the magma source can be carried. Gold and silver as well as many other precious metals can be moved by hydrothermal waters. As the water cools and its chemistry alters, it may dump these exotic ions. The less soluble unwanted burden, if sufficiently concentrated, may become the bonanza of some grizzled prospector or mining geologist. Water, precious enough in this sere region, is often the purveyor of treasured metals.

What to do with the ordinary dissolved load? Most streams and rivers carry their soluble loads to the sea. In the Great Basin, however, the rivers are isolated from the sea. Hence, the salts are temporarily stored in desert basins. Salts rim some of the playas, and limey deposits fill the pores of sands and gravels washed into the lowlands. The alkali flats and bitter springs that disappointed or killed the early emigrants attest to the soluble loads carried by Great Basin waters and the precipitation within the basins. The Great Basin, however, is an ephemeral storage bin. Eventually, rivers will cut the margins of the Great Basin or an arm of the ocean will pierce its interior, and the precipitated salts will again continue their journey to the sea.

Water, however, is not always so subtle as to utilize the silent patient dissolving of rock. Sometimes water uses the great range in desert temperature to attack the rock. Snowmelt or rainwater, filling fractures in the rock, will often freeze at night. The expansion of the growing ice crystals is a force too strong for even rock to resist. Every spring, trails and roads in the high country of the Great Basin are littered by rocks and boulders that have been shattered free by the freeze-thaw cycles of winter. The mountains come apart when there are few people there to watch the action.

Occasionally, even here in this parched region, the assault is even more frontal than that mounted by ice. My memory still reverberates with the roar of flash floods and the painful pounding on my unprotected skull by massive raindrops the size of quarters.

Using the force granted by gravity, and with large elevation differences between ranges and basins, water roars down mountain flanks toward the valleys below. This energy mechanically rips at rock and forces grains apart. This is no indirect patient pulling apart of ionic bonds. The rocks are torn apart physically and the fragments are tumbled downslope. The farther they move from the mountains, the smaller are the fragments carried by the fluid grist mill. Finally, in the center of the valleys, only the finest muds wash onto the plain. Through millennia these muds veneer the central valley and form playas. The flanks of the mountains, then, are graded. Large, coarse, angular blocks in the highlands grade downward like the sieve sizes at a gravel pit, through more rounded cobbles, to orbicular

sand grains, to the finest talcum-powder muds of the playas.

The slopes of the high mountains are steep, but along the flanks, buried by the detritus from above, the slope angle diminishes. Obeying simple constraints described by physics, the top surface of the fans trace precisely a catenary curve—the same curve followed by a broken telephone wire hanging limply from a pole to the ground. The steep end is supported by the mountain and the flat end is the playa. Playas are the most horizontal land surface to be found in the Great Basin. Following a rain, the playa may fill with water covering tens or hundreds of square miles—two inches deep. No islands or mudflats appear anywhere above the wavelets, and sandpipers casually wade across the entire lake.

Water has an amazing capacity to store heat, and it usually gives up this heat begrudgingly in nature. Seashores generally have mild climates because the sea absorbs the summer's heat, thus cooling the air. The ocean slowly releases the summer warmth during the winter, thus ameliorating winter's cold. There are incredible temperature variations every day in the Great Basin. The scarcity of water removes this thermal insurance from the landscape. Daytime temperatures soar over 100° F, and dusk forces the mercury down so fast that one can watch the thermometer fall. Diurnal temperature ranges in excess of 50° F are common. This is a daily spread in excess of the yearly range in many parts of earth.

Water, therefore, in its disguised role as the dissolver or in its more obvious role as the eroder is the primary agent of erosion in the Great Basin. Water defines the boundaries of the region geographically as well as the characteristics displayed by its rocks, climate, and life forms. Another paradox: here, where there is often so little water that lizards pant, it is water that is the creator and sculptor of the landscape and its life.

Surface Water

Surface water is certainly not common in the Great Basin, but virtually every mountain range has a few small streams that originate in the highlands from

Springs are amazingly abundant in this desert region. Only evaporated water leaves the Great Basin. All the pre- cipitation which infiltrates the ground must resurface again. *Stephen Trimble*.

snowmelt or summer rains. The basin floors may receive less than three inches of precipitation in a year. The high mountains are rain forests in comparison. Twenty or thirty inches of rain and snow fall on the highest ridges, and many of the mountains are blanketed by snow for half the year.

This abundance of high-altitude water disappears in several ways. Much evaporates into the parched atmosphere, some is used by the plants for transpiring, and the remainder infiltrates into the ground. Another water paradox: in this land of drought, abundant water lies stored in the subsurface. The gravels of most of the deep valleys are saturated res-

ervoirs, many of which contain more water than Lake Mead. The reservoirs, when filled to overflowing, leak the excess out to the surface.

Rarely, even in this desert region, are you more than five miles from an underground leakage. However, you must know where to look. Springs and seeps abound in almost every range, and even in parched Death Valley I have found water in most of the canyons draining the adjacent mountains.

There are also lakes in the Great Basin—desert lakes. When the underground reservoirs totally overflow with water, the reservoir level rises above the land surface and fills the valleys that have no outlets.

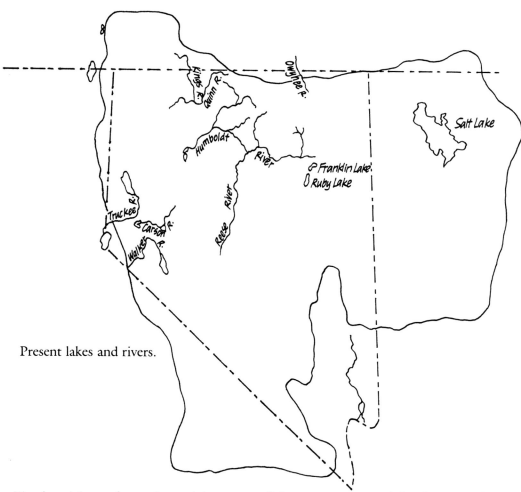

Present lakes and rivers.

To the visitor of our desert lakes, one of the most poignant memories is the strange anomaly of abundant water in shimmering blue lakes with shores so barren that there is often not even grass growing.

Pyramid, Walker, Pahranagat, Ruby, Franklin, Warner, Great Salt—the lakes of the Great Basin are merely residual puddles from the recent past. The onset of the ice ages created an environment of cooling and greater precipitation in the Great Basin. Snows blanketed the mountains and glaciers moved down many of the montane valleys. Summer meltwaters poured down the washes to the valleys. The infiltrating moisture filled the subsurface gravels to overflowing, and lakes filled many of the valleys. With no outlet to the sea, the Great Basin became

an inland sea, with the ranges standing above the water like north-south ships swinging at anchor before a strong north wind. Indeed, in those days a canoe would have been more useful to traverse the Great Basin than a stout pair of legs.

The Great Basin has expanded like a balloon during basin-and-range uplift in the past seventeen million years. The highest elevations within the basin are in the central portion. The eastern and western edges are rimmed by high plateaus or mountain ranges. The largest lakes are therefore trapped against the east-west borders. Lake Bonneville filled the eastern basin and Lake Lahontan covered the west. Leakage out of the margins of the region occurred then as now to the north by the Snake River

Water shapes and molds the individual grains, the boulders, and the mountains in this arid land. *John Running*.

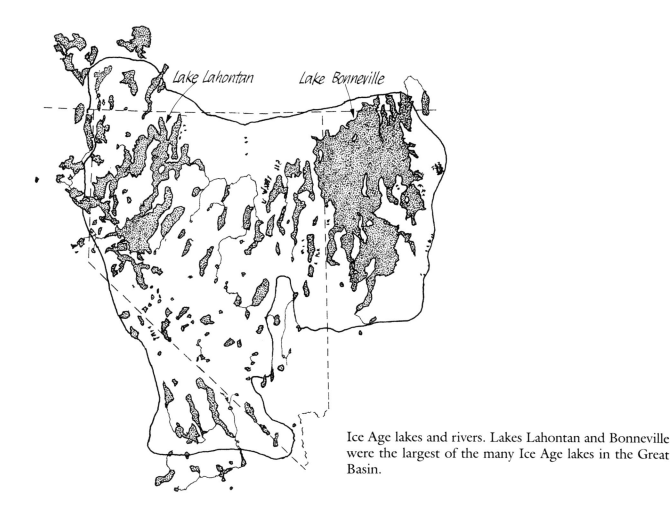

Ice Age lakes and rivers. Lakes Lahontan and Bonneville were the largest of the many Ice Age lakes in the Great Basin.

or to the south by the Colorado.

Today, lake shorelines are carved into most Great Basin valleys, but the reservoirs are now subsurface. There is too little precipitation. Great Salt Lake and Utah Lake remain as small remnants of the former Bonneville. Pyramid and Walker lakes, near the Sierra wall, remain from Lake Lahontan.

GROUNDWATER

The ancient Ice Age lakes are mostly gone. Beneath the desert valleys, protected from evaporation by the thick gravel roof of the reservoir, are the remnant waters of the colder times. These reservoirs are maintained at a full stage by inflowing subsurface waters and represent money in the bank for eco-

nomic use in the Great Basin. Mining, agriculture, and urban areas all draw on these reservoirs. Some reservoirs have undergone severe depletion due to overdrafting by wells. Like a bank account, withdrawals in excess of deposits leads to reduction of capital and eventual bankruptcy. With a sharp reduction in recharge since glacial times, it is easy to overdraw the watery accounts.

Wise management is essential. Hydrologic disaster results from the greed of overdrawing groundwater reservoirs. Springs dry up as the reservoir no longer spills its excess to the surface. The Devils Hole pupfish, an Ice Age fish that sought refuge in a spring in southern Nevada when the glacial lakes shriveled, was seriously threatened when the adjacent pumping of wells reduced the water level in the spring. Wells dry up as the water table sinks, and they must be deepened to chase the water downward. The cost of pumping the water increases as the depths increase. Water salinity in wells increases as the increased suction draws highly saline waters from the small spaces in shale and clay beds. The grains in the sediments settle and compact together as the water is withdrawn. Thus, the aquifer loses pore space and will never serve as effectively as a reservoir in the future. If compaction is extreme, the land surface may subside, cracking and destroying structures built on the sinking zones. Such subsidence has long been noted in Las Vegas Valley.

A balanced water withdrawal would seem to be the most logical method for long-term water use. If the withdrawals match the deposits, the steady-state condition will allow indefinite use. Some water-resource managers, however, argue for overexploitation. They give evidence of urban areas that heavily overdraw groundwater, allowing the population to expand. The large population can then afford to import water from more distant surface or subsurface water sources. The depleted groundwater basin can then be largely abandoned. They cite Las Vegas as a classic example of such water utilization. The groundwater reservoir was heavily overdrafted, but this cheap water allowed a burgeoning population to gain the economic advantage to pipe in water from the Colorado River.

The simplest groundwater flow systems move from ranges to the adjacent valleys. These local systems saturate the sediments in the valley. If there are no bedrock leaks below the gravels, the sediments fill with water and a series of springs or lakes will mark the discharge areas in the valley interior. Such areas often lack surface water but have green areas of tall sagebrush, tamarisk, or other plants that draw water from a shallow water table. These phreatophytes, or pump plants, are sensitive to water depths and salinity, and the depth to water and the water quality can be determined by knowing the phreatophytic species. A well drilled in the area of recharge where water is moving downward will not

have a flow to the surface, but wells drilled in the discharge portion of the flow system where water is moving upward may be artesian.

Not all the reservoirs fill to the surface. Many have subsurface leaks. Water pressure forces water through fractures or pore spaces in the bedrock. The mountain ranges between the valleys may act as effective barriers for surface water but not for underground movement of water. Infiltrating through the permeable rock, the water may leak through the range into the next lowest valley, creating regional flow systems in the subsurface.

The water in these systems can be markedly different. Local systems generally contain water that has been underground only a short time and only in the shallow subsurface. Such water is usually cool and has low salinity. It has not traveled deep enough to become geothermally warmed, nor has it been moving through rock long enough to become saturated with dissolved minerals. Also, since the source is local, variations in the amount of rainfall or snowpack will result in flow variations in springs. During dry years springs may dry up totally.

Regional water, on the contrary, is a long-distance voyager. Moving deep, it becomes warm or hot. Where the deep, hot water intersects faults or fractures it may short-circuit its flow and move to the surface. Hot springs, seeps, and geysers are common throughout the Great Basin. This geothermal water is often saturated with salts, either due to its passage through rocks with highly soluble minerals or because of its long residence time in rock. Some geologists believe there is also a significant contribution of salts to some thermal water from magmatic sources. Surface variations in rainfall are not significant to deep water. It flows at a steady rate year after year.

The central areas of some Great Basin valleys have large sandy dry lakes. These playas are among the flattest terrains known. Playas come in two types—wet or dry. Of course, after a rain any playa surface may be wet. But the basis of comparison should be made after a long period of drought. Those playas that overlie a leaky underground reservoir are dry. If

Thermal springs (from G. Waring, 1965).

the dry lake occupies a valley that has a full sub-surface reservoir, springs and seeps often keep the playa surface wet.

The recharge, or dry, playas will rarely be salty. Downward-moving waters will carry the salts out of the basin. Discharge playas, without hemorrhages in the internal plumbing, will fill with water. The water evaporates at the surface and concentrates the salts. Discharge playas often have saline lakes or muddy salt-encrusted playas. Tourists are often tempted to leave the blacktop highway while driving near a playa to cross the mudflat. It is wise to know the difference between recharge and discharge playas, since there are enterprising towing companies who make a good business near the discharge playas.

15 Caves

SPELUNKERS like to explore caves. Speleologists study them. Most of us are fascinated by dark recesses under the ground. The soluble Paleozoic carbonate rocks of the eastern Great Basin and the rapid changes in the water table through the coming and going of the Ice Age have created many beautiful solution caves in the area. Lava flows are sometimes hollow, and erosion often scours openings in volcanic rocks. More than three hundred caves have been described in the Great Basin, and more are still being found in the less-traveled recesses of isolated mountain ranges.

Early residents of the Great Basin used these caves as shelters. Ground sloths entered Gypsum Cave in southern Nevada more than eleven thousand years ago. Untidy housekeepers, they left piles of dung on the cave floor. Geologists, carefully examining the dung, can determine the plants eaten and partially digested by these giant sloths. These plants represent the vegetation in the cave vicinity during that Ice Age time. Indians later used this same cave and many others throughout the Great Basin. Early Anglo explorers, including Jedediah Strong Smith, sought shelter in these caves. Early newspapers often carried the exciting news of cave discoveries by sheepherders, cowboys, and prospectors. Often these were rediscoveries of caves long used by the resident Indians.

Speleologists, like other scientists, love to divide into threes. Great Basin caves are segregated into three types: corrosion caves, solution caves, and lava caves. Corrosion caves are the result of the physical erosion of an opening into rock. Solution caves are the result of the dissolving action of water. Lava caves are associated with volcanic flows.

Corrosion occurred from wave or current action along the shores of the Great Basin's many Ice Age lakes. The advances and recessions of lake levels, fluctuating with the vicissitudes of glacial regimen, scoured holes into the shores. Most of these corrosion caves are small and often contain Indian artifacts. They provided perfect campsites for Indians hunting or fishing along the shorelines. Rock paintings and carvings are found in or around the entrances to many such shelters.

Mechanical weathering forms another type of corrosion cave. Streams, frost wedging, or rock falls may leave holes or tumbled rocks that can be used as shelters. Fracturing or faulting often weakens rock for erosive forces. The resulting recess may be sufficiently large to hunker into during a winter storm. Our Great Basin rocks have had a long and violent geologic history of faulting, volcanism, and continental crushing. Consequently, mechanically weathered caves are common. During the past seventeen million years basin-and-range faulting and uplift have uplifted many of these broken and fractured rocks and exposed them to erosion. The presence of Indian artifacts demonstrates that these mechanical caves were often used as shelters for the Great Basin's earliest human inhabitants. They still serve the purpose admirably. I have many memories of sheltering from driving rainstorms or sudden wind-whipped blizzards in these ancient refuges.

Solution caves, however, make the most impressive underground caverns. Carbonate rocks, the limey excesses of Paleozoic seas, are abundant in the eastern Great Basin. This was the region of shelf limestones, and today these carbonates are dissolving. Rainwater is slightly acid, and with increased industrialization and burning of fossil fuels it is becoming increasingly so. Soil acids also acidify the infiltrating water. Moving along bedding planes, joints, or faults, the acid water dissolves the carbon-

Lehman Caves, Nevada. The concealing darkness of caves hides the slow processes of solution and deposition. Time is the prime ingredient in the formation of these stalactites. *John Running*.

ate and enlarges the openings until caves form. Most solution occurs underground, where the waters coalesce below the water table. Large caverns may form in this zone. As the water table lowered with the end of the Ice Age, the caverns emerged into air. Dripping waters leaking into the air-filled caves from above can form stalactites, stalagmites, or many other varieties of cave formations. Lehman Caves National Monument, high in the Snake Range of eastern Nevada, contains over ten thousand feet of cave passages that are beautifully adorned with cave formations. Timpanagos Cave National Monument in the Wasatch Range east of Provo, Utah, also is a beautiful solution cave protected by the National Park Service.

Some caves form within volcanic rocks. Molten lavas cool first along their margins—bottom, sides, and, eventually, top. The hot interior where flowage may still occur is insulated, as in a pipe or tube. If the lava source is cut off and the lava still flows from the end of the tube, a cave will result as the lava flows out the end of the tube. Such lava tubes are preserved in the Lava Beds National Monument in northern California. I know of no other experience in my life so totally stygian as that of exploring a lava tube. Caves are singularly dark to begin with, but couple this with uniformly black walls, floor, and roof. The blackness absorbs every ray of light from the cave lantern or flashlight. The light beam disappears as though into an astronomical black hole. You feel your way along the floor or walls and often cannot see the dark rock an inch in front of your dilated pupils. I have lumps on my skull to prove it.

Some caves have been intersected by miners following ore bodies. These caves may have formed from the original deposition of sulfides providing the cavity space. Sulfuric acid waters resulting from the oxidation and removal of sulfide ores may have further enlarged the original cavities. Stories, perhaps apocryphal, relate how miners have found (and since lost) caves lined with pure gold dust floors, surged by tidal water flows every day that have panned the gold from the sands.

Some unusual caves have been found along the shores of the Ice Age lakes in the tufa deposited by springs. Some of these may have formed in a manner opposite that of solution caves by deposition under lake water of limey tufa around mats of vegetation that have since rotted away, leaving a hole. In such a case, the hole was there first and was later enclosed by the limestone.

Caves sometimes form a unique ecosystem. The interior of caves is a delightful escape from summer heat or winter cold. Temperatures in caves are close to the average annual temperature of the region. Thus, they seem warm in the winter and cool in the summer. Bats, moths, crickets, scorpions, and millipedes find the constant cave environment an interesting place to live or rest. Pack rats also sometimes live in caves in such numbers or for so many generations that their urine and feces dry into a resinous substance that binds their nests into a hard black or dark-brown lump. Sometimes these rat middens contain vivid yellow and orange colors mixed with the black. Such colored bodies have been termed amberat. Some geologists, with notably steady stomachs, dissect these middens to determine the plant remains or spores and pollen. Since the middens can often be accurately dated by radioactive decay methods, they are useful guides to past climates.

Some Great Basin caves have been mindlessly destroyed. I vividly recall the thrill of exploring a cave near Las Vegas that was so rich in stalactites and stalagmites that passage through the cave to map its geology was difficult. Within a few years of its location being made known in a newspaper article, the cave was completely stripped of all its adornment. Now it is a littered, smooth-walled garbage pit. Since then, several other caves with excellent formations have been discovered near Las Vegas. Their entrances have been anonymously sealed with natural rubble to protect their interiors until official protection can be provided.

Red Rock Canyon Recreation Lands–Spring Mountain State Park. This scenic land of red and white Mesozoic sandstones lies just a few miles west of Las Vegas. Some of the most beautiful geologic formations and structures in the southern Great Basin may be seen in this area. *Tom Brownold*.

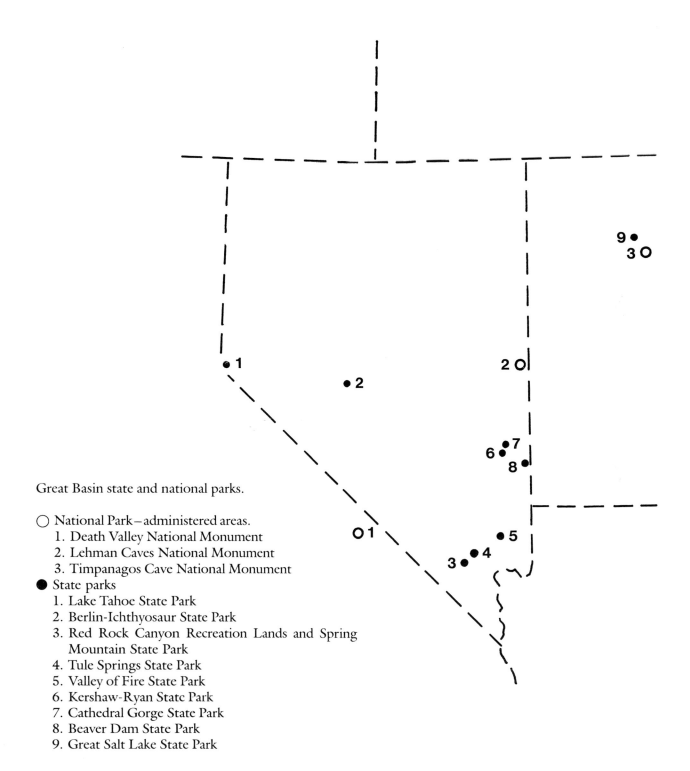

Great Basin state and national parks.

○ National Park—administered areas.
    1. Death Valley National Monument
    2. Lehman Caves National Monument
    3. Timpanagos Cave National Monument
● State parks
    1. Lake Tahoe State Park
    2. Berlin-Ichthyosaur State Park
    3. Red Rock Canyon Recreation Lands and Spring
       Mountain State Park
    4. Tule Springs State Park
    5. Valley of Fire State Park
    6. Kershaw-Ryan State Park
    7. Cathedral Gorge State Park
    8. Beaver Dam State Park
    9. Great Salt Lake State Park

Cathedral Gorge State Park. Geologically young lake sediments are eroded into a badlands topography. *Tom Brownold.*

mountains, seems to lie lifeless under the blazing desert sun. Names attribute a netherworld source for its features—Devil's Cornfield, Dante's View, Furnace Creek, and the Devil's Golfcourse. Names that penetrate the mind with frightful images. Names, however, reflect human perceptions. The plants, animals, and desert-adapted humans find this a delightful place to live or visit. Some are so content here that they would be unable or unwilling to live anywhere else. Here, too, the parched but gentle hand of Aeolus, the wind god, has added an artistic finishing touch to the landscape.

Fine sands and muds, washed down from surrounding mountains, lie unconsolidated and vulnerable on the valley floor. Desert winds have gently scooped out the loose detritus and shaped deflation hollows several feet deep. Roots of widely spaced arrow weed are exposed as the land surface is eroded down, and a strange landscape results. Stalks of dry vegetation stand in hourglass form with their tangled roots exposed above the desert floor—the Devil's Cornfield.

One of the fundamental concepts of physics is that matter is never created or destroyed. The aeolian sleight of hand, the removal of tons of desert soil from around the flanks of the arrow weed, must finally be reckoned. The transition from desert soil to the towering curvilinear sand piles seems almost magical. The sand dunes of Death Valley are one of the most intriguing and beguiling landscapes in the monument.

Desert winds moving up Death Valley, laden with silt and sand, are stalled and swirled by the air sinking into the valley through Emigrant Canyon and Furnace Creek. The air in the center of the gently swirling vortex is still, like the eye of a hurricane. Here the wind has no energy, and the sands swirling around the vortex are inexorably sucked into the center and dropped. Sand piles grow taller, vulnerable to aeolian forces.

As winds lift over a sand pile, air is compressed on the windward side of the dunes and the velocity of the wind increases. The increased velocity results in increased energy. The energetic winds on the windward side of the dune scour downward and move sand grains across the mound. On the far side of the dune, the air descends and relaxes. The less energetic wind drops the sand grains. The deposited sand rolls down the steep slope of the lee side, leaving layers inclined at a steep angle—the angle of repose, the steepest angle at which unenergetic sand lies. A sand dune, sliced by a curious geologist, a child, or by the whims of the wind, will have numerous steeply inclined layers in its interior. The inclinations will shift with time, depending upon the vagaries of wind direction, and form cross beds within the dune.

The sand grains, moved by the force of the winds, saltate over the dune surface. This bouncing and hopping action often pits the surface of the quartz grains and frosts them with a cloudy translucency. Each grain, viewed closely under a hand lens, resembles the frosted window glass in a bathroom.

Timpanagos Cave National Monument

The entrance to Timpanagos Cave is high on the walls of the canyon. Here is a cave that requires an uphill hike before the descent into the underground passages. The carbonate rocks of the Paleozoic were uplifted and cracked by earth's movements. The Wasatch Mountains were thrust high into the cooler, moisture-laden air. The water trickled down through the cracks and faults of the carbonates, enlarging them by solution and abrasion. Weirdly formed and shaped formations stretch the imagination of the spelunker.

STATE PARKS
Berlin-Ichthyosaur State Park

The Mesozoic was truly a fearsome time. Giant meat-eating reptiles searched the forests of cycads and ferns for anything to eat, living or dead. Muddy shorelines, covered with vegetation, were deeply indented with the footprints of the great carnivores. Giant dragonflies perched on the swaying tips of the horsetails. The oceans were no refuge. The sea surface was occasionally lashed into a white froth by huge marine sea lizards. These monsters, up to seventy feet long, had huge elongated jaws studded with conical teeth for grasping and piercing prey. Eyes the size of dinner plates assured vision in the murky depths of the sea. Powerful muscles drove a twenty-five-foot-long tail. A huge triangular fin, six feet across, tipped the tail. Large flippers extended laterally from the front and rear of the body to aid in high-speed turns. For about 135 million years, these monsters ranged the world's oceans and were the supreme marine predators—the ichthyosaurs.

Today, the ghost town of Berlin lies high on the western slope of the Shoshone Range. Sagebrush and pinyon pines grow around and through the dwellings of the miners. Some of the mining town's early residents lie interred in the nearby cemetery. Around the corner, up Union Canyon, are buried the remains of the earlier inhabitants—ichthyosaurs.

In the Late Triassic, two hundred million years ago, Berlin was a much different place than the desert miners knew. There were no high mountain ranges and no long desert valleys. A warm shallow sea lay sparkling in the tropical sun. A low land area rose to the east, with broad bordering mudflats and swamps. Sonomia had firmly accreted itself to the continent approximately coincident with the muddy shoreline and slowly foundered below the waves. Fine detritus, mud and sand, washed off the low-lying land. Lime precipitated out of the warm waters as a fine ooze and mixed with the muds. Logs washed seaward in streams and rivers, became waterlogged, and sank into the limey muds. They slowly petrified into limestone, the calcareous counterparts of the brightly colored siliceous agatized logs of Arizona's Petrified Forest.

This shallow mudflat became a deathtrap for marine creatures caught in the shallows by falling tides. Some animals with beautiful coiled shells, ammonites, were washed up on the shore. But the most dramatic sight would have been the great ichthyosaurs.

Here on the mud were stranded at least thirty-seven of the beasts. The sixty-foot reptiles lay like beached whales. Probably trapped alive by the falling tides, their huge bodies heaved as they gasped for breath in the hot equatorial sun. The long taper-ing snouts, studded by rows of pointed teeth, twisted into the mud beside their bodies during their death throes. The bodies rotted and many bones disarticulated. Flippers and vertebrae, bounded by tough ligaments, sometimes remained intact. As much as fifty tons of saurian protoplasm and bone disintegrated, most to rot and release elements back to the Triassic seas. But enough bones remained to reveal much of the lives of these ancient sea monsters. Buried by the constant influx of mud, the bones were petrified by the limey waters and preserved in the rock strata. Later geologic events piled thousands of feet of rock over the ancient bones. Lava flows and ash beds covered the great pile of sediments like a coffin lid.

During the formation of the Basin and Range, the ancient sediments were heaved upward along the faults that uplifted the Shoshone Range. Recent erosion scoured off the volcanic lid, exposing the underlying sediments and reopening the two-hundred-million-year-old burial site of the ichthyosaurs.

Berlin miners noticed the fossils. They made fireplaces out of the bones. Boys at the school in Union Canyon used spherical clam shells they called lizard heads as projectiles in their slingshots. The large bones were first identified in 1928 by Stanford geology professor Siemon Muller. The first large-scale excavations were begun in 1954 by Berkeley paleontologist Charles Camp. In 1970, the Nevada legislature protected the site as a state park and in 1977 passed a bill naming the ichthyosaur as our state fossil. It seems ironic that a desert, land-locked state would have a sea monster for its fossil.

A large building shelters the rock quarry where Camp and scores of youthful helpers disinterred the fossil remains. Today, you stare down at the two-hundred-million-year-old reptilian graveyard with awe, dwarfed both by the creatures and by time.

Valley of Fire State Park

Fifty-five miles from the bright lights of Las Vegas. The neon flashing and incandescent glowing is far away. Yet here, too, the air is electric. Here there is

Valley of Fire. Elephant Rock.
Donna Gripentog McKay

a glow of color and vibration. This land of color is illuminated by the rocks themselves. The air has substance, and the subtle shifts of light and shadow accentuate the texture of the rock. Like an amoeba, I sense everything around me. My hand is drawn to the rock. The rough cool surface refreshes my fingertips. The fragrance of the creosote bush is pungent, and the slightly choking odor of desert dust catches deep in my throat.

The desert of the southern Great Basin is direct—blunt and honest in its reality—and nowhere is this more obvious than in the Valley of Fire. This is a place of exciting, eerie beauty, with fantastic forms created by the scouring of uplifted land. At once forbidding and attracting, these forms are among nature's finest sculpting.

The red colors are derived from rust. Iron oxidized in the sun. The rocks seem to trap the color of the setting sun within their own texture. The sand grains that compose the dramatic red rocks were blown by Jurassic winds. Heaped into piles in that long-ago desert, the dunes were preserved by burial and later cementation by groundwater. The arcuate symmetry of the innards of fossil dunes are preserved in the rocks as gigantic crossbeds.

Some hulking gray mountains rim the edges of the park. Uplifted by vertical basin-and-range faults, these ranges expose the limestones laid down along the Paleozoic shelf in that ancient south equatorial sea. The rocks have been stacked, as though by a blackjack dealer, into huge sheets of carbonate rocks. This is the Overthrust Belt. The pressures from titanic forces have shoved rocks over rocks and stacked them into geologic puzzles. The normal superposition of both rocks and fossil assemblages have been shuffled into confusion.

Here, shortly after the turn of the century, the great Yale geologist Chester Longwell scratched his head over these confused ranges. His first trips to these mountains were afoot, leading a burro. Then he used a Model T. One of his last trips was by helicopter. Eventually, he sorted them out. His monumental work on the Muddy Mountains still stands as a guide to another generation of space-age geologists. The importance of long hours of pressing your nose against a rock, sweating in the southern Nevada sunshine, and listening to the rocks speak will never be replaced by the LandSats or electron probe. New tools are helpful, but nothing can replace the intimate contact with the rock itself. In these hills, Longwell showed us the way.

The handiwork of erosion is particularly evident here. Fractured rocks, crossbeds, differing textures—all become the materials to be molded by erosional carving. This is a landscape of wild beauty. Elephants, ducks, fish, and even pregnant bears seem to be carved out of solid stone. Holes beckon for children to explore. Adults who have not outlived their youth scramble up sloping sandstone rock to see what lies on the other side.

For centuries people have rested in the cool summer shade of a cliff or snuggled into a hollow to shelter from the arctic blasts of winter. The apparent harshness of this landscape belies the comforts an experienced desert dweller can find. Petroglyphs were chipped through the black desert varnish by the early Indian residents. This ancient stone art serves as a link from past to present. I thrill to place my hand over the stone outline of another hand, placed on the same rock thousands of years ago.

Red Rock Canyon Recreation Lands and Spring Mountain State Park

The Keystone Thrust Fault. Geologists come from all over the United States to see this feature. Remember the lesson of superposition? Old rocks underlie the young. The basement is built before the first floor. Here lies the exception to the rule. The old gray Paleozoic limestones are thrust over the younger red rocks of the Jurassic. This is no mild disconformity. There are hundreds of millions of years of discrepancy. The breaking of the rule is so flagrantly and beautifully illustrated by the sharp color contrast that many introductory geology textbooks have a photograph of this feature. It was described by Chester Longwell and has since been studied, mapped, and photographed by thousands of geologists.

The Wilson Cliffs rise almost vertically above the floor of Red Rock Canyon. They can easily be seen from Las Vegas, standing as a wall to the west of the city. The tan cliffs are horizontally slashed with red. The ancient dunes are highlighted by rust streaks. Mesozoic. Capping the sandstone of the cliffs is gray, tree-covered limestone. Paleozoic.

This is one of the finest examples of the Overthrust Belt to be seen anywhere. The pressures of continent formation have thrust ancient rocks up and over the young. The old limestones of Pahrump Valley to the west have been shoved easterly and now peer somberly down from the cliff crest into Las Vegas Valley.

The sandstones fractured during thrusting and uplift. Recent snowmelt and rain runoff follow the weaknesses of the rock. What were once fractures are now eroded into deep canyons that penetrate westerly into the escarpment of the Wilson Cliffs. A hike into the shadowed recess of the canyons reveals a microworld of watery beauty. Toads cling to sandstone walls, and ferns six feet long shadow pools of deep, cool water. Canyon wrens create echoing waterfalls of clear whistling sound from niches in the cliffs. Red Rock is a delightful escape from the juxtaposed urban area. Here you can be immersed in the sounds and sights of nature and rest in the shade of a thrust fault.

Cathedral Gorge State Park

Remember sitting on a beach or in a mud puddle as a youth and letting sand or mud filter through your fingers? With a little care, you learned to build spires and castles out of the liquid sand drops. Cathedral Gorge seems like a king-sized child's fantasyland.

Lake muds and thin limestones settled slowly onto Tertiary lake bottoms. The fine detritus eroded from uplands was trapped until hundreds of feet of sediment accumulated. There were few changes in the lake regime or in the type of incoming sediment. Uniformity.

Uplift has exposed the long-buried lake sediments today. Erosion has sculpted the soft uniform

Cathedral Gorge.
Donna Gripentog McKay

layers into a fairyland. Spires, cathedrals, mounds, domes—all covered with polygonal mudcracks—lie bleached and lifeless in the sun. Rain washes down into cracks in the muds and slices vertical cuts into sides of the forms. Some are isolated by fluvial washing and others are vertically etched. Still others are penetrated by sheer slices only a few feet wide. Children and adventurous adults slide into these crevices and deeply penetrate the inner parts of the old lake muds. Peering up to the sky through the narrow slits gives you a feeling of kinship to the fossil interred in the rock. You become one with the rock.

Beaver Dam and Kershaw-Ryan State Parks

Tertiary volcanic rocks dominate both of these Nevada parks. A large caldera collapsed in the Caliente area following the expulsion of vast quantities of ejecta. The cliffs and erosional landforms in these parks are carved from ash-fall tuffs and lava flows associated with this intense period of volcanism.

Tule Springs (Floyd R. Lamb) State Park

Warm waters emerge in large springs along the axis of the desert floor. These springs are roughly coincident with the Las Vegas Valley Shear Zone, and the groundwater may be finding access to the surface along these faults. Tule Springs makes a perfect oasis in the center of the parched Las Vegas Valley.

White beds of mud surround the park. These are the remnants of Ice Age lakes that formerly filled the central parts of the valley. These beds have been carefully studied for faunal remains. They are rich in Ice Age creatures from small freshwater snails to the bones of mammoths. The Las Vegas Valley is not thought to have had a single large body of water, as many of the more northern Great Basin valleys did. There was probably a series of discontinuous lakes connected by perennial streams. Giant beavers may have dammed the flow and formed new lakes. Perhaps pine trees grew along the shores and green vegetation filled the water-rich valley. Which would have been more startling to the Paiutes of two hundred years ago—looking backward to the Pleistocene greenery and water, or forward to the bright lights and urban excesses of modern Las Vegas?

Great Salt Lake State Park

We usually think of the Great Basin as parched desert lands. Paradoxically, the largest lake west of the Mississippi River lies in the eastern Great Basin. The Great Salt Lake covers more than 1,700 square miles. A moist eastern state, Rhode Island, could be totally submerged beneath the waters of this desert lake.

The unique quality of many Great Basin lakes is their lack of outlet to the sea. The Great Salt Lake is the bottom of the hill. Any minerals washed into the lake by streams draining the Wasatch Range to the east or by flash floods coursing eastward from the Nevada-Utah deserts are all concentrated in Great Salt Lake. The water can leave by evaporation but the salts are fated by nature's laws to remain behind. More than two million tons of minerals are carried into the lake every year. There are an estimated five billion tons of salts now in the lake. Only the Dead Sea in the Middle East is more salty. The lake is sometimes eight times as salty as the ocean. Your body becomes so buoyant in such dense water that it is impossible to dive.

The lake constantly changes level relative to changes in rainfall and evaporation. Recently, cool wet winters have caused a major expansion of the lake.

The lake is so saline that it doesn't freeze in the winter. The high salinity also absorbs solar heat and the lake remains relatively warm through the winter. The air over the lake rises, due to thermal convection, and carries the moisture-laden air to cold heights. Here the moisture condenses and snow falls. This lake effect can cause heavy snowfall in Salt Lake City.

Lake Tahoe, Nevada, State Park

The paradoxes of Great Basin lakes are nowhere more evident than in this beautiful alpine park. Here, straddling the California-Nevada border on the western margin of the droughtlands of the Great Basin, is the second deepest lake in the United States (Crater Lake, Oregon, is deeper). The maximum water depth of Lake Tahoe is 1,645 feet. The surface area of the lake is 193 square miles. The total water volume contained in the lake is enormous, due to its depth. The 126 million acre feet of water is enough to cover the entire Great Basin to a depth of three-quarters of a foot.

In a land characterized by high heat, Tahoe's surface water averages 68° F in summer and 50° F in the winter. At a depth of two hundred feet, the water maintains a constant temperature of 41° F. It is so free of sediment and the cold waters are so resistant to algae that its water is among the clearest in the world.

Lake Tahoe is not a typical Great Basin lake. It has an outlet, although not to the sea. It lies in a large downdropped block of crust. A large fault dropped down in response to tensional forces across this part of the Sierra. Later lava flows and ash falls, together with glacial debris from the Ice Age, have blocked the natural drainage from the north end of the downfaulted block. A man-made dam creates a few extra vertical feet for water storage. This addition amounts to a considerable enlargement of storage capacity in the lake.

17 The Future

THE GEOLOGIC HISTORY of the Great Basin relates tales of long-ago volcanism and mountain building. Events are described in the millions, hundreds of millions, or even billions of years. Such a tale tends to leave the reader with a false illusion of security. Monumental events have occurred in our backyards, but they were long, long ago. Surely such catastrophic occurrences as orogenies and volcanism must be finished, neatly segregated by geologists into the pages and diagrams of reports and books.

But no! Quite the contrary. There is clear evidence that mountain-building forces and volcanic fires still stir today beneath the seemingly solid earth under our feet. The earth is active and geology is a dynamic science.

The western border of the Great Basin is a zone of especially active seismicity. This belt follows the eastern front of the Sierra Nevada and includes Reno and Carson City. On March 26, 1872, California's Owens Valley was shaken by a great earthquake. The eastern side of the Sierra Nevada lurched upward, relative to the valley, by about thirteen feet and slid laterally to the northwest by about sixteen feet. The surface was ruptured by the fault for a distance of about one hundred miles. At Lone Pine, twenty-three of the 250 inhabitants were killed and fifty-two of the fifty-nine adobe houses were destroyed.

The earthquake was felt from Reno to San Diego.

A swarm of earthquakes has been recorded in the past several years near Mammoth Lakes, California. These tremors are thought to be related to the movement of magma in the shallow subsurface that could be a precursor to volcanism along the eastern flank of the Sierra.

Recent faulting has torn the fabric of Nevada's surface as well. In 1869, faulting ruptured the land at Olinghouse. Later breaks occurred in 1903 at Wonder, in 1915 at Pleasant Valley, in 1932 at Cedar Mountain, in 1934 at Excelsior Mountain, and in 1954 at Dixie Valley–Fairview Peak and Rainbow Mountain. These earthquakes generated both extensional and strike-slip displacements.

One of the most hazardous zones for earthquakes lies along the eastern boundary of the Great Basin. The Wasatch front was created by vertical forces uplifting the Wasatch Range. These forces are still active today, and the risk from earthquake along the base of the Wasatch is very high.

Seismic potential in regions that have little or no historic seismicity is frequently underestimated. This is especially true for the western United States, where the historical record is only about one century long. To understand the future, it is necessary to evaluate the past, say historians. As geology is a historical science, surely the same adage is true.

Red Rock Canyon. Crossbeds of Mesozoic dunes are etched into relief today by the slow processes of weathering and erosion. *John Running*.

Acknowledgments

THOUSANDS of prospectors, miners, gem collectors, and geologists have, for more than one hundred years, been picking and poking in the ranges and sandy valleys of the Great Basin. They are the ones who have gathered the geological information about the Great Basin. My role was to piece together that material.

Specific individuals, however, were responsible for the creation of the book. Rick Stetter—editor par excellence, river runner, and field companion—helped in a thousand ways from the beginning to the end. Grant money from the Fleischmann Foundation (given to the University of Nevada Press for their Great Basin Natural History Series) funded my research and field work. The geological review of the manuscript was done by John Stewart of the U.S. Geological Survey and by J. V. Tingley of the University of Nevada, Reno. They provided excellent suggestions and help. Eugene Smith and Diane Pyper-Smith reviewed a chapter. Bridget Boulton, a geological editor, made many suggestions of a grammatical and geological nature. Mary Hill patiently and accurately copy-edited the manuscript and proofread the galleys. Donna Gripentog McKay researched the references and also supplied artwork. My good friend Nate Stout drew the many illustrations. Brad Fiero conducted library research. Mike Stowers and Lynn Dryer, University of Nevada, Las Vegas, Audio Visual Services, were a great help with the word processing during the writing and editing stages. Photographers John Running and Tom Brownold were excellent field companions and artists. Additional photography was supplied by Stephen Trimble, Tony Diebold, the U.S. Geological Survey, and the Nevada Bureau of Mines. David Comstock was responsible for the layout and design of the book, as well as for hours of enjoyable work and discussions together. Cameron Sutherland of the University Press helped pull together the many last-minute details.

It is to my family—Ellen, Brad, Scott, and Kim—that I owe the greatest thanks. Their understanding and love gave me the opportunity and support to research and write this book.

References

Armentrout, J. M., M. R. Cole, and H. TerBest, Jr. (eds.). 1979. *Cenozoic paleogeography of the western United States*. Society of Economic Paleontologists and Mineralogists, Pacific Section, Pacific Coast Paleogeography Symposium 3.

Burchfiel, B. C. 1975. Nature and controls of Cordilleran orogenesis, western United States: Extensions of an earlier synthesis. *American Journal of Science* 272: 97–118.

Coney, P. J. 1979. Tertiary evolution of Cordilleran metamorphic core complexes. Pp. 15–28 in *Cenozoic paleogeography of the western United States*, ed. J. M. Armentrout, M. R. Cole, and H. TerBest, Jr.

Dickinson, W. R. 1977. Paleozoic plate tectonics and the evolution of the Cordilleran continental margin. Pp. 137–155 in *Paleozoic paleogeography of the western United States*, ed. J. H. Stewart, C. H. Stevens, and A. E. Fritsche.

———. 1979. Cenozoic plate tectonic setting of the Cordilleran region in the United States. Pp. 1–13 in *Cenozoic paleogeography of the western United States*, ed. J. M. Armentrout, M. R. Cole, and H. TerBest, Jr.

Hamilton, W. 1978. Mesozoic tectonics of the western United States. Pp. 23–70 in *Mesozoic paleogeography of the western United States*, ed. D. G. Howell and K. A. McDougall.

Hill, M. 1975. *Geology of the Sierra Nevada*. California Natural History Guide 37. Berkeley: University of California Press.

Hintze, L. F. 1973. *Geologic history of Utah*. Brigham Young University Geology Studies, vol. 20, pt. 3. Studies for Students no. 8.

Howell, D. G., and K. A. McDougall (eds.). 1978. *Mesozoic paleogeography of the western United States*. Society of Economic Paleontologists and Mineralogists, Pacific Section, Pacific Coast Paleogeography Symposium 2.

Johnson, J. G., and A. Pendergast. 1981. Timing and mode of emplacement of the Roberts Mountains allochthon, Antler orogeny. *Geological Society of America Bulletin* 92: 648–658.

Langenheim, R. L., Jr., and E. R. Larson. 1973. *Correlation of Great Basin stratigraphic units*. Nevada Bureau of Mines and Geology Bulletin 72.

Longwell, C. R., E. H. Pampeyan, B. Bowyer, and F. J. Roberts. 1965. *Geology and mineral deposits of Clark County, Nevada*. Nevada Bureau of Mines and Geology Bulletin 62.

Lucchitta, I. 1972. Early history of the Colorado River in the Basin and Range province. *Geological Society of America Bulletin* 83: 1933–1948.

McKee, E. H. 1971. Tertiary igneous chronology of the Great Basin of the western United States—Implications for tectonic models. *Geological Society of America Bulletin* 82: 3497–3502.

Newman, G. W., and H. D. Goode (eds.). 1979. Basin and Range symposium and Great Basin field conference. Rocky Mountain Association of Geologists and Utah Geological Association.

Nilsen, T. H., and J. H. Stewart. 1980. The Antler orogeny—Mid-Paleozoic tectonism in western North America. *Geology* 8: 298–302.

Schweickert, R. A. 1981. Tectonic evolution of the Sierra Nevada Range. Pp. 87–131 in *The geotectonic evolution of California*, ed. W. G. Ernst. Englewood Cliffs, N.J.: Prentice-Hall.

Smith, R. B., and G. P. Eaton (eds.). 1978. *Cenozoic tectonics and regional geophysics of the western Cordillera*. Geological Society of America Memoir 152.

Speed, R. C. 1971. Golconda thrust, western Nevada—Regional extent. Pp. 199–200 in Geological Society of America Abstracts with Programs, vol. 3, no. 2.

———. 1979. Collided Paleozoic microplate in the western United States. *Journal of Geology* 87: 279–292.

Stewart, J. H. 1971. Basin and Range structure—A system of horsts and grabens produced by deep-seated extension. *Geological Society of America Bulletin* 82: 1019–1043.

———. 1976. Late Precambrian evolution of North America: Plate tectonics implication. *Geology* 4: 11–15.

———. 1980. *Geology of Nevada, a discussion to accompany the geologic map of Nevada.* Nevada Bureau of Mines and Geology Special Publication 4.

———, C. H. Stevens, and A. E. Fritsche (eds.). 1977. *Paleozoic paleogeography of the western United States.* Society of Economic Paleontologists and Mineralogists, Pacific Section, Pacific Coast Paleogeography Symposium 1.

Wright, L. A. 1976. Late Cenozoic fault patterns and stress fields in the Great Basin and westward displacement of the Sierra Nevada block. *Geology* 4: 489–494.

———, and B. W. Troxel. 1970. Summary of regional evidence for right-lateral displacement in the western Great Basin. (Discussion of paper by J. H. Stewart, J. P. Albers, and F. G. Poole, 1968.) *Geological Society of America Bulletin* 81: 2167–2173.

Index